DNA Computing Models

Zoya Ignatova · Israel Martínez-Pérez ·
Karl-Heinz Zimmermann

DNA Computing Models

 Springer

Zoya Ignatova
Cellular Biochemistry
Max Planck Institute of Biochemistry
82152 Martinsried by Munich
Germany
ignatova@biochem.mpg.de

Karl-Heinz Zimmermann
Institute of Computer Technology
Hamburg University of Technology
21071 Hamburg
Germany
k.zimmermann@tuhh.de

Israel Martínez-Pérez
Institute of Computer Technology
Hamburg University of Technology
21071 Hamburg
Germany
martinez-perez@tuhh.de

ISBN 978-1-4419-4464-1   e-ISBN: 978-0-387-73637-2
DOI: 10.1007/978-0-387-73637-2

Cover illustration:

Printed on acid-free paper

9 8 7 6 5 4 3 2 1

springer.com

Preface

Biomolecular computing was invented by Leonard Adleman, who made head-lines in 1994 demonstrating that DNA – the double-stranded helical molecule that holds life's genetic code – could be used to carry out computations. DNA computing takes advantage of DNA or related molecules for storing information and biotechnological operations for manipulating this information. A DNA computer has extremely dense information storage capacity, provides tremendous parallelism, and exhibits extraordinary energy efficiency. Biomolecular computing has an enormous potential for in vitro analysis of DNA, assembly of nanostructures, and in vivo calculations.

The aim of this book is to introduce the beginner to DNA computing, an emerging field of nanotechnology based on the hybridization of DNA molecules. The book grew out of a research cooperation between the authors and a graduate-level course and several seminars in the master's program in Computer Engineering taught by the third author at the Hamburg University of Technology during the last few years. The book is also accessible to advanced undergraduate students and practitioners in computer science, while students, researchers, and practitioners with background in life science may feel the need to catch up on some undergraduate computer science and mathematics. The book can be used as a text for a two-hour course on DNA computing with emphasis on mathematical modelling.

This book is designed not as a comprehensive reference work, but rather as a broad selective textbook. The first two chapters form a self-contained introduction to the foundations of DNA computing: theoretical computer science and molecular biology. Chapter 2 concisely describes the abstract, logical, and mathematical aspects of computing. Chapter 3 briefly summarizes basic terms and principles of the transfer of the genetic information in living cells. The remaining chapters contain material that for the most part has not previously appeared in textbook form. Chapter 4 addresses the problem of word design for DNA computing. Proper word design is crucial in order to successfully conduct DNA computations. Chapter 5 surveys the first generation of DNA computing. The DNA models are laboratory-scaled and human-operated, and basically aim at solving complex computational problems. Chapter 6 addresses the second generation of DNA computing. The DNA models are molecular-scaled, autonomous, and partially programmable, and essentially target the in vitro analysis or synthesis of DNA. Chapter 7 is devoted to the newest generation of DNA computing. The DNA models mainly aim at performing logical calculations under constraints found in living cells.

We have not tried to trace the full history of the subjects treated – this is beyond our scope. However, we have assigned credits to the sources that are as readable as possible for one knowing what is written here. A good systematic reference for the material covered are the Proceedings of the Annual International Workshop on DNA Based Computers.

First of all, we would like to thank Professor Volker Kasche and Professor Rudi Müller for valuable support and for providing laboratory facilities for our experimental work. We are grateful to Dr. Boris Galunsky, Stefan Goltz, Margaret Parks, and Svetlana Torgasin for proofreading, and we express our thanks to Wolfgang Brandt and Stefan Just for technical support. Finally, we thank our students for their attention, their stimulating questions, and their dedicated work, in particular, Atil Akkoyun, Gopinandan Chekrigari, Zhang Gong, Sezin Nargül, Lena Sandmann, Oliver Scharrenberg, Tina Stehr, Benjamin Thielmann, Ming Wei, and Michael Wild.

Hamburg, Munich   Zoya Ignatova
December, 2007   Israel Martínez-Pérez
                                                                  Karl-Heinz Zimmermann

Contents

Acronyms

Mathematical Notation

$\emptyset$	empty set
$\mathbb{N}$	set of natural numbers
$\mathbb{N}_0$	set of non-negative integers
$\mathbb{Z}$	set of integers
$\mathbb{R}$	set of real numbers
$\mathbb{R}_0^+$	set of non-negative real numbers
$\mathbb{R}^+$	set of positive real numbers
$P(S)$	power set of a set S
δ_{ij}	Kronecker delta
$\circ$	function composition
$\lfloor x \rfloor$	largest integer $\leq x$
$\lceil x \rceil$	least integer $\geq x$
ϵ	empty string
Σ	alphabet
Δ	DNA alphabet
Σ^n	set of all length-n strings over Σ
Σ^*	set of all strings over Σ
Σ^+	set of all non-empty strings over Σ
$\Sigma^\bullet$	set of all circular strings over Σ
$\Sigma^\star$	set of all signed strings over Σ
$|x|$	length of string x
x^R	mirror image of string x
x^C	complement of string x
x^{RC}	reverse complement of string x
$\bar{x}$	reverse complement of string x
$\bullet x$	circular word
$\bar{f}$	negation of Boolean function f
$|M|$	size of automaton M

$L(M)$	language of automaton M
$L(G)$	language of grammar G
$\mathbb{B}$	set $\{0,1\}$
$\mathbb{F}_n$	nth Boolean algebra
nt	number of nucleotides
bp	number of base pairs
aa	number of amino acids
$\Delta G°$	Gibbs free energy
$\Delta H°$	enthalpy
$\Delta S°$	entropy
T	temperature
T_m	melting temperature
$[X]$	concentration of reactant X
V	volume
d_H	Hamming distance
d_H^ϕ	ϕ-Hamming distance
σ_λ	similarity function
σ_β	block similarity function
$\mathrm{com}_U(G)$	U-complement of G
$\mathrm{loc}_v(G)$	local complement of G at v

Physical Units

Angstrom	$1\,\text{Å} = 10^{-10}\,\text{m}$
Atomic mass	$1\,\text{Da} = 1.661 \cdot 10^{-27}\,\text{kg}$
Avogadro number	$N_A = 6.022 \cdot 10^{23}\,1/\text{mol}$
Boltzmann constant	$k_B = 1.38 \cdot 10^{-23}\,\text{J/K}$
Dielectric constant of vacuum	$\epsilon_0 = 8.854 \cdot 10^{-12}\,\text{F/m}$
Dipole moment	$1\,\text{D} = 3.34 \cdot 10^{-30}\,\text{Cm}$
Electron charge	$e = 1.602 \cdot 10^{-19}\,\text{C}$
Electron mass	$m_e = 9.109 \cdot 10^{-31}\,\text{kg}$
Gas constant	$R = 1.987\,\text{cal/(K mol)}$
Planck constant	$h = 6.626 \cdot 10^{-34}\,\text{Js}$
Reduced Planck constant	$\hbar = h/(2\pi)\,\text{Js}$
Mole	$1\,\text{mol} = 6.022 \cdot 10^{23}\,\text{molecules}$
Molarity	$1\,\text{M} = 6.022 \cdot 10^{23}\,\text{mol/l}$

Chemical Notation

H	hydrogen atom
O	oxygen atom
C	carbon atom
N	nitrogen atom
S	sulfur atom
P	phosphor atom
A	adenine
C	cytosine
G	guanine
T	thymine
U	uracil

Chapter 1
Introduction

Abstract This introductory chapter envisions DNA computing from the perspective of molecular information technology, which is brought into focus by three confluent research directions. First, the size of semiconductor devices approaches the scale of large macromolecules. Second, the enviable computational capabilities of living organisms are increasingly traced to molecular mechanisms. Third, techniques for engineering molecular control structures into living cells start to emerge.

Nanotechnology

Nanotechnology focuses on the design, synthesis, characterization, and application of materials and devices at the nanoscale. Nanotechnology comprises near-term and molecular nanotechnology. Near-term nanotechnology aims at developing new materials and devices taking advantage of the properties operating at the nanoscale. For instance, nanolithography is a top-down technique aiming at fabricating nanometer-scale structures. The most common nanolithography technique is electron-beam-directed-write (EBDW) lithography in which a beam of electrons is used to generate a pattern on a surface.

Molecular nanotechnology aims at building materials and devices with atomic precision by using a molecular machine system. Nobel Prize-winner R. Feynman in 1959 was the first who pointed towards molecular manufacturing in his talk "There's plenty of room at the bottom," in which he discussed the prospect of maneuvering things around atom by atom without violating physical laws. The term nanonechnology was coined by N. Taniguchi in 1974, while in the 1980s E. Drexler popularized the modelling and design of nanomachines, emphasizing the constraints of precision, parsimony, and controllability, performing tasks with minimum effort. Eric Drexler's nanomachines include nano-scale manipulators to build objects atom by atom, bearings and axles built of diamond-like lattices of carbon, waterwheel-like pumps

Z. Ignatova et al., *DNA Computing Models*,
DOI: 10.1007/978-0-387-73637-2_1, © Springer Science+Business Media, LLC 2008

to extract and purify molecules, and tiny computers with moving parts whose size is within atomic scale.

Nanotechnology relies on the fact that material at the nanoscale exhibits quantum phenomena, which yield some extraordinary bonuses. This is due to the effects of quantum confinement that take place when the material size becomes comparable to the de Broglie wavelength of the carries (electrons and holes behaving as positively charged particles), leading to discrete energy levels. For instance, quantum dots are semiconductors at the nanoscale consisting of 100 to 100,000 atoms. Quantum dots confine the motion of (conduction band) electrons and (valency band) holes in all three spatial directions. Quantum dots are particularly useful for optical applications due to their theoretically high quantum yield (i.e., the efficiency with which absorbed light produces some effect). When a quantum dot is excited, the smaller the dot, the higher the energy and intensity of its emitted light. These optical features make quantum dots useful in biotechnological developments as well. Recently, D. Lidke and colleagues (2004) successfully employed quantum dots to visualize the movement of individual receptors on the surface of living cells with unmatched spatial and temporal resolution.

Biotechnology

Modern biotechnology in the strong sense refers to recombinant DNA technology, the engineering technology for bio-nanotechnology. Recombinant DNA technology allows the manipulation of the genetic information of the genome of a living cell. It facilitates the alteration of bio-nanomachines within the living cells and leads to genetically modified organisms. Manipulation of DNA mimics the horizonal gene transfer (HGT) in the test tube.

HGT played a major role in bacterial evolution. It is thought to be a significant technique to confer drug-resistant genes. Common mechanisms for HGT between bacterial cells are transformation, the genetic alteration of a cell resulting from introducing foreign gene material, transduction, in which genetic material is introduced via bacterial viruses (bacteriophages), and bacterial conjugation, which enables transfer of genetic material via cell-to-cell contact. HGT appears to have some significance for unicellular eukaryotes, especially for protists, while its prevalence and importance in the evolution of multicellular eukaryotes remains unclear. Today, the HGT mechanisms are used to alter the genome of an organism by exposing cells to fragments of foreign DNA encoding desirable genes, including those from another species. This DNA can be either transiently internalized into the cell or integrated into the recipient's chromosomes. Thus, it can be replicated and inherited like any other part of the genome. HGT holds promising applications in health care and in industrial and environmental processing.

Bio-Nanotechnology

Nanotechnology was invented more than three billion years ago. Indeed, nanoscale manipulators for building molecule-sized objects were required in the earliest living cells. Today, many working examples of bio-nanomachines exist within living cells. Cells contain molecular computers, which recognize the concentration of surrounding molecules and compute the proper functional output. Cells also host a large collection of molecule-selective pumps that import ions, amino acids, sugars, vitamins and all of the other nutrients needed for living. By evolutionary search and modification over trillions of generations, living organisms have perfected a plethora of molecular machines, structures, and processes (Fig. 1.1).

Bio-nanomachines are the same size as the nanomachines that are designed today. But they hardly resemble the machines of our macroscopic world and they are less familiar than E. Drexler's manipulators built along familiar rigid, rectilinear designs. D. Goodsell recently claimed that the organic, flexible forms of bio-nanomachines can only be understood by looking at the forces that made possible the evolution of life. The process of evolution by natural selection places strong constraints on biological molecules, their structure and their function. As a consequence of the evolution of life, all living organisms on earth are made of four basic molecular building blocks: proteins, nucleic acids, polysaccharides, and lipids. Proteins and nucleic acids are built in modular form by stringing subunits (monomers) together based on genetic information. These polymers may be formed in any size and with

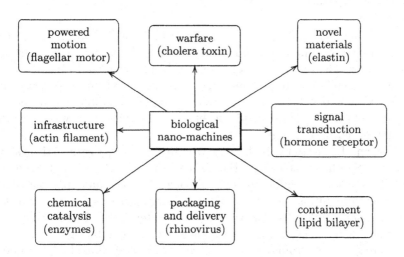

Fig. 1.1 Molecular bio-nanomachines in living cells.

monomers in any order so that they are remarkably flexible in structure and function. On the other hand, lipids and polysaccharides are built by dedicated bio-machines. Each type of new lipid or polysaccharide requires an entirely new suite of synthetic machines. Consequently, lipids and polysaccharides are less diverse in structure and more limited in function than proteins are.

The principles of protein structure and function may yield insight into nanotechnological design and fabrication. Proteins are synthesized in a modular and information-driven manner by the translation machinery of the cell, and the design of proteins is limited by a dedicated modular plan given by the genetic code. Proteins can aggregate in larger complexes due to errors in the protein-synthetic machinery or changes in the environmental conditions, so the size of proteins that may be consistently synthesized is limited. These aggregates can be built accurately and economically by protein-protein interactions based on many weak interactions (hydrogen bonds) and highly complementary shapes of interacting surfaces. Proteins are synthesized in cells and are transported to their ultimate destinations or diffuse freely in a crowded collection of competitors. A typical protein will come into partial contact with many other types of proteins and must be able to discriminate its unique target from all others. Proteins constantly flex at physiological temperatures, with covalent bonds remaining connected, and reshaped hydrogen bonds and salt bridges linking portions of the molecule or aggregate. Proteins even breathe, switching between different conformations and allowing atoms or small molecules to pass.

Synthetic Biology

The term synthetic biology was introduced by E. Kool and other speakers at the annual meeting of the American Chemical Society in 2000. Synthetic biology in broader terms aims at recreating the properties of living systems in unnatural chemical systems. That means, assembling chemical systems from unnatural components so that the systems support Darwinian evolution and are thus biological. Thus, synthetic biology may provide a way to better understand natural biology.

DNA and RNA are the molecular structures that support genetic systems on earth. Synthetic biology partially shows that the DNA and RNA backbone is not a simple scaffold to hold nucleobases but has an important role in molecular recognition, and the repeating charge provides the universal feature of genetic molecules that they work in water. Recently, S. Benner and coworkers (2003) constituted a synthetic genetic system by eight nucleotides that were generated from the natural nucleobases by shuffling hydrogen-bond donating and accepting groups. This system is part of the Bayer VERSANT branched DNA diagnostic assay which provides a reliable method to quantify HIV-1 RNA in human plasma.

Molecular Self-Assembly

Molecular self-assembly is an autonomous process of nanofabrication in which molecules or aggregates are formed without the influence of an outside source. The physicist H.R. Crane (1950) provided two basic design concepts required for molecular self-assembly. First, the contact or combined spots on the components must be multiple and weak. Thus, an array of many weak interactions is considered preferable to a few very strong interactions because the latter may lead to interactions with wrong candidates. Second, the assembled components must be highly complementary in their geometrical arrangement so that tightly packed aggregates can result. These two concepts can be observed in numerous protein-protein structures, as already mentioned.

Molecular self-assembly can theoretically create a wide range of aggregates. However, a major inherent difficulty is that the exact set of components and interactions that will construct the aggregate is difficult to determine. Recent advances in biotechnology and nanotechnology provided tools necessary to consider engineering at the molecular level. DNA computation introduced by L. Adleman in 1994 blazed a trail for the experimental study of programmable biochemical reactions, the self-assembly of DNA structures.

DNA Nanotechnology

DNA nanotechnology was initiated by N. Seeman in the 1980s. It makes use of the specificity of Watson-Crick base pairing and other DNA properties to make novel structures out of DNA. The techniques used are also employed by DNA computing and thus DNA nanotechnology overlaps with DNA computing. A key goal of DNA nanotechnology is to construct periodic arrays in two and three dimensions. For this, DNA branched junctions with specific sticky ends are designed that self-assemble to stick figures whose edges are double-stranded DNA. Today, this technology provides cubes, truncated octahedrons, and two-dimensional periodic arrays, while three-dimensional periodic arrays are still lacking. One ultimate goal is the rational synthesis of DNA cages that can host guest molecules whose structure is sought by crystallography. This would overcome the weakness of the current crystallization protocol and provide a good handle on the crystallization of all biological molecules.

Computing

A digital computer can be viewed as a network of digital components such as logic gates. The network consists of a finite number of components and the components can take on a few states. Thus, the network has only a finite number of states, and hence any realizable digital computer is a finite state machine, although with a vast number of states. Today, these machines are realized by digital electronic circuits mainly relying on transistor technology. The success of digital electronic circuits is based on low signal-to-noise ratio, inter-connectability, low production costs, and low power dissipation. Digital electronic circuits scaled predictably during the last 30 years, with unchanged device structure and operability. Another decade of scaling appears to be feasible.

Digital computers excel in many areas of applications, while other interesting information processing problems are out of reach. The limitations are of both a theoretical and physical nature. Theoretical limitions are due to the nature of computations. The first model of effective computation was introduced by the Turing machine, which is essentially a finite state machine with an unlimited memory. In view of the generally accepted Church's thesis, the model of computation provided by the Turing machine is equivalent to any other formulation of effective computation. A machine capable of carrying out any computation is called a universal machine. Universal Turing machines exist, and every personal computer is a finite-state approximation of a univeral machine. A general result in computability reveals the existence of problems that cannot be computed by a universal machine despite potentially unlimited resources. Efficient computations can be carried out on practical computers in polynomial time and space. However, there are computational problems that can be performed in exponential time and it is unknown whether they can be performed in polynomial time and space. A prototype example is the travelling salesman problem that seeks to find a route of minimal length through all cities in a road map.

Biomolecular Computing

Current attempts to implement molecular computing fall into two categories. In the first are studies to derive molecular devices that mimic components of conventional computing devices. Examples are transistors from carbon-based semiconductors and molecular logic gates. The second includes investigations to find new computing paradigms that exploit the specific characteristics of molecules. Examples that fall into this category are computions based on diffusion-reaction or self-assembly.

A physical computation in a digital computer evolves over time. Information is stored in registers and other media, while information is processed by using digital circuits. In biomolecular computing, information is stored by biomolecules and processing of information takes place by manipulating biomolecules. The concept of biomolecular computing was theoretically discussed by T. Head in 1987, but L. Adleman in 1994 was the first to solve a small instance of the travelling salesman problem with DNA. Adleman's experiment attracted considerable interest from researchers hoping that the massive parallelization of DNA molecules would one day be the basis to outperform electronic computers, when it comes to the computation of complex combinatorial problems. However, soon thereafter, researchers realized some of the drawbacks related to this incipient technology: a growing number of error-prone, time-consuming operations, and exponential growth of DNA volume with respect to problem size. Although some new concepts like molecular self-assembly counteracted these difficulties, no satisfactory solution to these problems has been found so far questioning the feasibility of this technology for solving intractable problems.

Therefore, molecular computing should not be viewed as a competitor for conventional computing, but as a platform for new applications. Progress in molecular computing will depend on both novel computing concepts and innovative materials. The goal of molecular information processing is to find computing paradigms capable of exploiting the specific characteristics of molecules rather than requiring the molecules to conform to a given specific formal specification.

References

1. Adleman LM (1994) Molecular computation of solutions of combinatorial problems. Science 266:1021–1023
2. Benner SA, Sismour AM (2005) Synthetic biology. Nature Rev Genetics 6: 533–543
3. Crane HR (1950) Principles and problems of biological growth. Sci Monthly 70:376–389
4. Carbone A, Seeman NC (2004) Molecular tiling and DNA self-assembly. LNCS 2340:219–240
5. Drexler KE (1992) Nanosystems, molecular machines, manufacturing and computation. Wiley and Sons, New York
6. Feynman RP (1961) Miniaturization. In: Gilbert DH (ed.) Reinhold, New York
7. Geyer C, Battersby T, Benner SA (2003) Nucleobase pairing in expanded Watson-Crick-like genetic information systems. Structure 11:1485–1498
8. Goodsell DS (2000) Biotechnology and nanotechnology. Sci Amer 88:230–237
9. Head T (1987) Formal language theory and DNA: an analysis of the generative capacity of specific recombination behaviors. Bull Math Biol 47:737–759
10. Kendrew J (1998) Encyclopedia of molecular biology. Blackwell Sci, Oxford
11. Lidke DS, Nagy P, Heintzmann R, Arndt-Jovin DJ, Post JN, Grecco H, Jares-Erijman EA, Jovin TM (2004) Quantum dot ligands provide new insights into erbB/HER receptor-mediated signal transduction. Nat Biotech 22:198–203
12. Leavitt D (2006) The man who knew too much: Alan Turing and the invention of the computer. Norton, London
13. Rawls R (2000) Synthetic biology makes its debut. Chem Eng News 78:49–53
14. Seeman N (1982) Nucleic acid junctions and lattices. J Theor Biol 99:237–247
15. Taniguchi N (1974) On the basic concept of nanotechnology. Proc Intl Conf Prod Eng Tokyo, Japan Soc Prec Eng
16. Wu R, Grossman L, Moldave K (1980) Recombinant DNA. Vol 68 Academic Press New York

Chapter 2
Theoretical Computer Science

Abstract This chapter provides a self-contained introduction to a collection of topics in computer science that focusses on the abstract, logical, and mathematical aspects of computing. First, mathematical structures called graphs are described that are used to model pairwise relations between objects from a certain collection. Second, abstract machines with a finite number of states called finite state automata are detailed. Third, mathematical models of computation are studied and their relationships to formal grammars are explained. Fourth, combinatorial logic is introduced, which describes logic circuits whose output is a pure function of the present input only. Finally, the degrees of complexity to solve a problem on a computer are outlined.

2.1 Graphs

Graph theory provides important tools to tackle complex problems in different parts of science.

2.1.1 Basic Notions

A *graph* is a pair $G = (V, E)$, consisting of a non-empty set V and a set E of two-element subsets of V. The elements of V are called *vertices* and the elements of E are termed *edges*. An edge $e = \{u, v\}$ is also written as $e = uv$ (or $e = vu$). If $e = uv$ is an edge, then u and v are *incident* with e, u and v are *adjacent*, and u and v form the *end-vertices* of e. In the following, we consider *finite* graphs (i.e., graphs with finite vertex sets). The number of vertices and edges of a graph G is called the *order* and *size* of G, respectively.

A graph is described by a *diagram*, in which the vertices are points in the drawing plane and the edges are line segments.

Z. Ignatova et al., *DNA Computing Models*,
DOI: 10.1007/978-0-387-73637-2_2, © Springer Science+Business Media, LLC 2008

Fig. 2.1 Diagram of the
graph in Example 2.1.

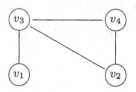

Example 2.1. The graph G with vertex set $V = \{v_1, \ldots, v_4\}$ and edge set $E = \{v_1 v_3, v_2 v_3, v_2 v_4, v_3 v_4\}$ is given by the diagram in Figure 2.1. $\Diamond$

A graph $G = (V, E)$ has neither *loops* nor *multiple edges*. Loops are one-element subsets of V (i.e., edges incidenting with only one vertex). Multiple edges are multisets over the two-element subsets of V. A *multiset* over a set M is a mapping $f : M \to \mathbb{N}_0$ assigning to each element m in M the number of occurrences $f(m)$ in the multiset.

Let $G = (V, E)$ be a graph. The number of edges which are incident with a vertex $v \in V$ is called the *degree* of v and is denoted by $d(v)$. A vertex v in G is called *isolated* if $d(v) = 0$. If all vertices in G have the same degree k, then the graph G is called *k-regular*.

Lemma 2.2. (Handshaking) *For each graph* $G = (V, E)$,

$$\sum_{v \in V} d(v) = 2|E| \, . \tag{2.1}$$

Proof. On the left hand side, each edge in the sum is counted twice, once for each vertex. $\square$

Corollary 2.3. *In each graph, the number of vertices of odd degree is even.*

Example 2.4. Can 333 phones be connected so that each phone is connected with three phones? The answer is no, because the sum of degrees in this network would be odd $(333 \cdot 3)$, contradicting the handshaking lemma. $\Diamond$

The *degree sequence* of a graph G is given by the decreasing list of degrees of all vertices in G. For instance, the graph in Figure 2.1 has the degree sequence $(3, 2, 2, 1)$. On the other hand, not every decreasing sequence of natural numbers is the degree sequence of a graph, such as $(5, 3, 2, 2, 2, 1)$, since the sum of degrees is odd.

Subgraphs

Let $G = (V, E)$ be a graph. A *subgraph* of G is a graph $G' = (V', E')$ with $V' \subseteq V$ and $E' \subseteq E \cap \binom{V'}{2}$, where $\binom{V'}{2}$ is the set of 2-element subsets of V'. The subgraph G' is considered to be *induced* from its edge set E'. If $E' = E \cap \binom{V'}{2}$, the subgraph G' is *induced* from its vertex set V'.

Fig. 2.2 Two subgraphs, G_1 and G_2, of the graph G in Figure 2.1.

Example 2.5. In view of the graph G in Figure 2.1, two of its subgraphs G_1 and G_2 are illustrated in Figure 2.2. The subgraph G_2 is induced from the vertex set $\{v_2, v_3, v_4\}$, while the subgraph G_1 is not because the edge v_2v_3 is missing.   ◇

Isomorphisms

Let $G = (V, E)$ and $G' = (V', E')$ be graphs. A mapping $\phi : V \to V'$ is called an *isomorphism* from G onto G', if ϕ is bijective and for all vertices $u, v \in V$, $uv \in E$ if and only if $\phi(u)\phi(v) \in E'$. Two graphs G and G' are termed *isomorphic* if there is an isomorphism from G onto G'. Clearly, isomorphic graphs have the same order, size, and degree sequence.

Example 2.6. The graphs in Figure 2.3 are isomorphic. An isomorphism is given by $\phi(v_i) = u_i$ for $1 \leq i \leq 4$.   ◇

2.1.2 Paths and Cycles

Let $G = (V, E)$ be a graph. A sequence $W = (v_0, \dots, v_k)$ of vertices $v_i \in V$ is called a *path* in G, if for each i, $1 \leq i \leq k$, we have $v_{i-1}v_i \in E$. The vertex v_0 is the *initial vertex* and the vertex v_k the *final vertex* in W. The *length* of W equals n, the number of edges in W. A path W is called *simple* if W contains each vertex at most once.

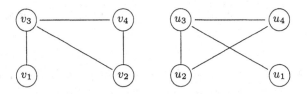

Fig. 2.3 Two isomorphic graphs.

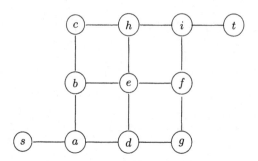

Fig. 2.4 A Manhattan network.

Example 2.7. The graph in Figure 2.4 contains several simple paths of length 6 such as (s, a, d, g, f, i, t) and (s, a, b, e, h, i, t). ◇

A *cycle* in G is a path in G, in which the initial and final vertex are identical. A cycle is called *simple* if it contains each vertex at most once (apart from the initial and final vertex). Each edge uv provides a simple cycle (u, v, u) of length 2.

Example 2.8. The graph in Figure 2.4 contains several simple cycles of length 6 such as (a, b, c, h, e, d, a) and (a, b, e, f, g, d, a). ◇

Connectedness

Let $G = (V, E)$ be a graph. Two vertices $u, v \in V$ are called *connected* in G, briefly $u \equiv_G v$, if $u = v$ or there is a path from u to v in G. If any two vertices in G are connected, then G is termed *connected*. For each vertex v in G, define the set of vertices connected to v as $C_G(v) = \{u \in V \mid u \equiv_G v\}$.

Theorem 2.9. *Let $G = (V, E)$ be a graph. The set of connected sets $C_G(v)$, $v \in V$, of G is a partition of V (i.e., the sets are non-empty and their union provides the overall set V, and any two sets are either equal or disjoint).*

A subgraph induced by a connected set of G is called a *component* of G. If G is connected, then there is only one component.

Example 2.10. The graph in Figure 2.5 consists of two components: $\{a, b\}$ and $\{c, d, e\}$. ◇

Theorem 2.11. *Let $G = (V, E)$ be a connected graph and let K be a simple cycle in G. If an edge $e \in G$ on the cycle K is deleted from G, the resulting subgraph of G is still connected.*

Fig. 2.5 A graph with
two components.

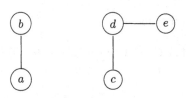

2.1.3 Closures and Paths

A *directed graph* is a pair $G = (V, E)$, consisting of a non-empty set V and
a subset E of $V \times V$ (Fig. 5.1). Each undirected graph can be assigned a
directed graph so that each edge $e = uv$ is replaced by the edges (u, v) and
(v, u). The edge set of a directed graph forms a binary relation on V. The
indegree of a vertex v in G is the number of incoming edges (u, v), $u \in V$,
and the *outdegree* of v is the number of outcoming edges (v, w), $w \in V$.

Let R be a binary relation on a set A (i.e., $R \subseteq A \times A$). Define the powers
of R inductively as follows:

- $R^0 = \{(a, a) \mid a \in A\}$,
- $R^{n+1} = R \circ R^n = \{(a, c) \mid (a, b) \in R, (b, c) \in R^n, b \in A\}$ for all $n \geq 0$.

Clearly $R^1 = R^0 \circ R = R$. The definition implies the following:

Theorem 2.12. *Let $G = (V, E)$ be a directed graph and let $n \geq 0$ be an
integer. The nth power E^n provides all paths of length n between any two
vertices in G.*

Define $R^+ = \bigcup_{n \geq 1} R^n$ and $R^* = \bigcup_{n \geq 0} R^n = R^+ \cup R^0$.

Theorem 2.13. *Let R be a binary relation on a set A. The relation R^+ is
the smallest transitive relation containing R. The relation R^* is the smallest
reflexive, transitive relation that contains R.*

Proof. Let $R' = \bigcup_{n \geq 1} R^n$. Claim that R' is transitive. Indeed, let $a, b, c \in A$
with $(a, b) \in R'$ and $(b, c) \in R'$. Then there are non-negative integers m
and n so that $(a, b) \in R^m$ and $(b, c) \in R^n$. Thus, $(a, c) \in R^{m+n}$ and hence
$(a, c) \in R'$. Moreover, $R = R^1$ and so $R \subseteq R'$.

Finally, let R'' be a transitive relation on A, which contains R. Claim
that $R' \subseteq R''$. Indeed, let $a, b \in A$ with $(a, b) \in R'$. Then there is a non-
negative integer n so that $(a, b) \in R^n$. Consequently, there are elements
$a_1, \ldots, a_{n-1}$ in A so that $(a, a_1) \in R$, $(a_i, a_{i+1}) \in R$ for all $1 \leq i \leq n-2$, and
$(a_{n-1}, b) \in R$. But R is a subset of R'' and so $(a, a_1) \in R''$, $(a_i, a_{i+1}) \in R''$
for all $1 \leq i \leq n-2$, and $(a_{n-1}, b) \in R''$. As R'' is transitive, it follows that
$(a, b) \in R''$ and so the claim is established.

The second assertion is similarly proved.   □

The relation R^+ is called the *transitive closure* of R, while the relation R^* is
termed the *reflexive, transitive closure* of R.

Distances

Let $G = (V, E)$ be a graph and let $u, v \in V$. Define the *distance* between u and v in G as follows:

$$d_G(u, v) = \begin{cases} 0 & \text{if } u = v, \\ \infty & \text{if } u \text{ and } v \text{ are not connected,} \\ l & \text{if } l \text{ is the length of a shortest path in } G \text{ from } u \text{ to } v. \end{cases} \quad (2.2)$$

Theorem 2.14. *Let $G = (V, E)$ be a graph. The distance d_G defines a metric on G. That is, for all $u, v, w \in V$, $d_G(u, v) = 0$ if and only if $u = v$, $d_G(u, v) = d_G(v, u)$, and $d_G(u, w) \leq d_G(u, v) + d_G(v, w)$.*

Notice that each metric d_G satisfies $d_G(u, v) \geq 0$ for all $u, v \in V$, because $0 = d_G(u, u) \leq d_G(u, v) + d_G(v, u) = 2d_G(u, v)$.

2.1.4 Trees

A graph is called *cycle-free* or a *forest* if it contains no simple cycles of length at least 3. A connected forest is called a *tree* (Fig. 2.6).

Theorem 2.15. *Each tree contains at least two vertices of degree 1.*

Proof. Let G be a tree. Let u und v be vertices in G so that their distance $d_G(u, v)$ is maximal. Let $W = (u, v_1, \ldots, v_{k-1}, v)$ be a shortest path in G from u to v. Suppose that u has two adjacent vertices, v_1 and w. Then by hypothesis, $d_G(w, v) \leq d_G(u, v)$. Thus there is a shortest path from w to v not using u. So G contains a simple cycle of length at least 3. A contradiction. Consequently, u has degree 1 and, by symmetry, also v. $\qquad\square$

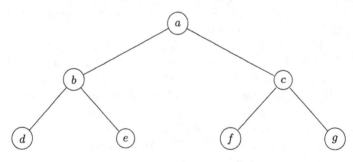

Fig. 2.6 A tree.

Theorem 2.16. *For each tree $G = (V, E)$, we have $|E| = |V| - 1$.*

Proof. The case $|V| = 1$ is clear. Let G be a tree with $|V| > 1$ vertices. In view of Theorem 2.15, the graph G contains a vertex of degree 1. If this vertex is deleted, the resulting subgraph $G' = (V', E')$ of G is a tree, too. By induction hypothesis, $1 = |V'| - |E'| = (|V| - 1) - (|E| - 1) = |V| - |E|$. □

Let $G = (V, E)$ be a graph. A *spanning tree* of G is a subgraph of G, which forms a tree and contains each vertex of G (Fig. 2.7).

Theorem 2.17. *Each connected graph contains a spanning tree.*

Proof. Let $G = (V, E)$ be a connected graph. If $|E| = 1$, then the assertion is clear. Let $|E| > 1$. If G is a tree, then G is its own spanning tree. Otherwise, there is a simple cycle of length at least 3 in G. Delete one edge from this cycle. The resulting subgraph G' of G has $|E| - 1$ edges and is connected by Theorem 2.11. Thus by induction hypothesis, G' has a spanning tree, and this spanning tree is also a spanning tree of G. □

Theorem 2.18. *A connected graph $G = (V, E)$ is a tree if and only if $|E| = |V| - 1$.*

Proof. Let $|E| = |V| - 1$. Suppose G is not a tree. Then G contains a simple cycle of length at least 3. Delete one edge from this cycle. The resulting subgraph $G' = (V, E')$ of G is connected by Theorem 2.11. The edge set in G' fulfills $|E'| < |V| - 1$. On the other hand, Theorems 2.16 and 2.17 imply that G contains a spanning tree with $|V| - 1$ edges, which lies in E'. A contradiction. The reverse assertion was proved in Theorem 2.16. □

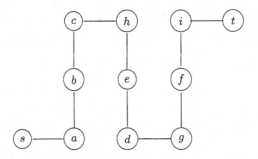

Fig. 2.7 A spanning tree of the graph in Figure 2.4.

Fig. 2.8 A bipartite graph with partition $\{\{a,b,c\},\{d,e,f\}\}$.

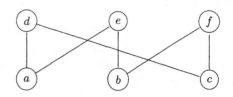

2.1.5 Bipartite Graphs

A graph $G = (V,E)$ is called *bipartite* if there is a partition of V into subsets V_1 und V_2 so that every edge in G has one end-vertex in V_1 and one end-vertex in V_2 (Fig. 2.8).

Theorem 2.19. *A connected graph G is bipartite if and only if G contains no cycles of odd length.*

Proof. Let $G = (V,E)$ be a bipartite graph with partition $\{V_1, V_2\}$. Let $K = (v_0, v_1, \ldots, v_k)$ be a cycle in G. If $v_0 \in V_1$, then $v_1 \in V_2$, $v_2 \in V_1$, and so on. Thus, $v_k = v_0 \in V_1$ and hence the cycle K has even length. If $v_0 \in V_2$, then the result is the same.

Conversely, assume that G contains no cycles of odd length. Let $v \in V$ and define

$$V_1 = \{u \in V \mid d_G(v,u) \equiv 1 \bmod 2\}$$

and

$$V_2 = \{u \in V \mid d_G(v,u) \equiv 0 \bmod 2\} \,.$$

Clearly, $\{V_1, V_2\}$ is a partition of V. Suppose that there is an edge uw in G with $u, w \in V_1$. Then there is a cycle, consisting of the edge uw, a path of length $d_G(w,v)$ from w to v, and a path of length $d_G(v,u)$ from v to u. This cycle has total length $1 + d_G(w,v) + d_G(v,u)$, which is odd by definition of V_1 and V_2. A contradiction. Similarly, there exists no edge uw in G with $u, w \in V_2$.   □

2.2 Finite State Automata

Finite state automata are a simple type of machine studied first in the 1940s and 1950s. These automata were originally proposed to model brain functions. Today, finite state automata are mainly used to specify various kinds of hardware and software components.

2.2.1 Strings and Languages

Let Σ be a finite set and let n be a non-negative integer. A *word* or *string* of length n over Σ is a sequence $x = a_1 \ldots a_n$ so that $a_i \in \Sigma$ for each $1 \leq i \leq n$. The *length* of a string x is denoted by $|x|$. The set Σ is termed *alphabet* and the elements of Σ are called *characters* or *symbols*. The *empty string* corresponds to the empty sequence and is denoted by ε. For instance, the strings of length at most 2 over $\Sigma = \{a, b\}$ are ε, a, b, aa, ab, ba, and bb.

Define Σ^n as the set of all strings of length n over Σ. In particular, $\Sigma^0 = \{\varepsilon\}$ and $\Sigma^1 = \Sigma$. Moreover, let Σ^* be the set of all strings over Σ (i.e., Σ^* is the disjoint union of all sets Σ^n, $n \geq 0$). Write Σ^+ for the set of all non-empty strings over Σ (i.e., Σ^+ is the disjoint union of all sets Σ^n, $n \geq 1$). Any subset of Σ^* is called a *(formal) language* over Σ.

The *concatenation* of two strings x and y is the string xy formed by joining x and y. Thus, the concatenation of the strings "home" and "work" is the string "homework". Let x be a string over Σ. A *prefix* of x is a string u over Σ so that $x = uv$ for some string v over Σ. Similarly, a *postfix* of x is a string v over Σ so that $x = uv$ for some string u over Σ.

A *monoid* is a set M which is closed under an associative binary operation, denoted by '$\cdot$', and has an *identity element* $\varepsilon \in M$. That is, for all x, y, and z in M, $(x \cdot y) \cdot z = x \cdot (y \cdot z)$, and $x \cdot \varepsilon = x = \varepsilon \cdot x$. This monoid is written as a triple $(M, \cdot, \varepsilon)$. In particular, the set Σ^* forms a monoid with the operation of concatenation of strings and with the empty string as the identity element. For any two languages L_1 and L_2 over Σ, write $L_1 L_2 = \{xy \mid x \in L_1, y \in L_2\}$ to denote their *concatenation*.

Let $(M, \cdot, \varepsilon)$ and $(M', \circ, \varepsilon')$ be monoids. A *homomorphism* from M to M' is a mapping $\phi : M \to M'$ so that for all $x, y \in M$, $\phi(x \cdot y) = \phi(x) \circ \phi(y)$ and $\phi(\varepsilon) = \varepsilon'$. An *anti-homomorphism* from M to M' is a mapping $\phi : M \to M'$ so that for all $x, y \in M$, $\phi(x \cdot y) = \phi(y) \circ \phi(x)$ and $\phi(\varepsilon) = \varepsilon'$. A homomorphism $\phi : M \to M$ is called a *morphic involution* if ϕ^2 is the identity mapping. The simplest morphic involution is the identity mapping. An anti-homomorphism $\phi : M \to M$ so that ϕ^2 is the identity mapping is termed an *anti-morphic involution*.

Let Σ be an alphabet. Each mapping $f : \Sigma \to \Sigma$ can be extended to a homomorphism $\phi : \Sigma^* \to \Sigma^*$ so that $\phi(a) = f(a)$ for each $a \in \Sigma$. To see this, put $\phi(a_1 \ldots a_n) = f(a_1) \ldots f(a_n)$ for each string $a_1 \ldots a_n \in \Sigma^*$. Similarly, each mapping $f : \Sigma \to \Sigma$ can be extended to an anti-homomorphism $\phi : \Sigma^* \to \Sigma^*$. For this, define $\phi(a_1 \ldots a_n) = f(a_n) \ldots f(a_1)$ for each string $a_1 \ldots a_n \in \Sigma^*$.

Single strands of DNA are quaternary strings over the DNA alphabet $\Delta = \{A, C, G, T\}$. Strands of DNA are oriented (e.g., AACG is distinct from GCAA). An orientation is introduced by declaring that a DNA string begins with the 5'-end and ends with the 3'-end. For example, the strands AACG and GCAA are denoted by 5'-AACG-3' and 5'-GCAA-3', respectively. Furthermore, in nature DNA is predominantly double-stranded. Each natural strand occurs with its reverse complement, with reversal denoting that the sequences of the two strands are oppositely oriented, relative to one other, and with complementarity denoting that the allowed pairings of letters, opposing one another on the two strands, are the *Watson-Crick pairs* $\{A, T\}$ and $\{G, C\}$. A double strand results from joining reverse complementary strands in opposite orientations:

$$5'-AACGTC-3'$$
$$3'-TTGCAG-5' \, .$$

DNA strands that differ by orientation are mapped onto each other by the *mirror involution* $\mu : \Delta^* \to \Delta^*$, which is the anti-homomorphism extending the identity mapping. For example, $\mu(AACG) = GCAA$. The mirror image of a DNA string x is denoted by $x^R = \mu(x)$. Moreover, the *complementarity involution* is the morphic involution $\phi : \Delta^* \to \Delta^*$ that extends the *complementarity mapping* $f : \Delta \to \Delta$ given by $f(A) = T$, $f(C) = G$, $f(G) = C$, and $f(T) = A$. For example, $\phi(AACG) = TTGC$. The complementary image of a DNA string x is denoted by $x^C = \phi(x)$. Finally, reverse complementary strands are obtained by the *reverse complementarity involution* or *Watson-Crick involution* $\tau = \mu\phi \, (= \phi\mu)$, which is composed of the mirror involution μ and the complementarity involution ϕ (in any order). For example, $\tau(AACG) = CGTT$. The reverse complementary image of a DNA string x is denoted by $x^{RC} = \tau(x)$.

2.2.2 Deterministic Finite State Automata

A finite state automaton can be thought of as a processing unit reading an input string and accepting or rejecting it. A *(deterministic) finite state automaton* is a quintuple $M = (\Sigma, S, \delta, s_0, F)$ so that Σ is an alphabet, S is a finite set of *states* with $S \cap \Sigma = \emptyset$, $s_0 \in S$ is the *initial state*, $F \subseteq S$ is the set of *final states*, and $\delta : S \times \Sigma \to S$ is the *transition function*, where the transition $\delta(s, a) = s'$ is also graphically written as $s \xrightarrow{a} s'$. The *size* of a finite state automaton M, denoted by $|M|$, is the number $|S| + |\delta|$.

Example 2.20. Consider the finite automaton M with state set $S = \{s_0, s_1\}$, input alphabet $\Sigma = \{a, b\}$, initial state s_0, final state set $F = \{s_0\}$, and transition function δ given by the transition graph in Figure 2.9. $\quad\diamond$

A finite state automaton M computes a string $x = a_1 \ldots a_n$ as follows: M starts in the initial state s_0, reads the first symbol a_1 and enters the state $s_1 = \delta(s_0, a_1)$. Then it reads the next symbol a_2 and enters the state $s_2 = \delta(s_1, a_2)$ and so on. After reading the last symbol a_n, the automaton enters the state $s_n = \delta(s_{n-1}, a_n)$. Therefore, the processing of an input string x can be traced by the associated path $(s_0, \ldots, s_n)$ in the transition graph. If the last state s_n is a final state, then M *accepts* the string x; otherwise,

Fig. 2.9 Transition
graph of finite state
automaton.

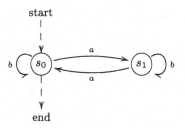

M rejects the string x. The *language* of M is the set of all strings accepted by M,

$$L(M) = \{x \in \Sigma^* \mid M \text{ accepts } x\} . \tag{2.3}$$

The multi-step behavior of a finite state automaton M can be formally described by the *extended transition function* $\delta^* : S \times \Sigma^* \to S$, which is inductively defined as follows:

- $\delta^*(s, \varepsilon) = s$,
- $\delta^*(s, ax) = \delta^*(\delta(s, a), x)$ for all $s \in S$, $a \in \Sigma$, and $x \in \Sigma^*$.

In particular, $\delta^*(s, a) = \delta(s, a)$ for all $s \in S$ and $a \in \Sigma$. The language of M is thus given by

$$L(M) = \{x \in \Sigma^* \mid \delta^*(s_0, x) \in F\} . \tag{2.4}$$

If $L = L(M)$ is a finite language, the size of the accepting automaton M is in the worst case proportional to the total length of all strings in L.

Example 2.21. Consider the finite state automaton M in Example 2.20. The language of M consists of all strings over Σ which contain an even number of a's. For instance, $\delta^*(s_0, abab) = s_0$ and $\delta^*(s_0, bbab) = s_1$.   ◇

2.2.3 Non-Deterministic Finite State Automata

Non-deterministic machines may provide several next states for each pair of state and input symbol. A *non-deterministic finite state automaton* is a quintuple $M = (\Sigma, S, \delta, S_0, F)$ so that Σ is an alphabet, S is a finite set of *states* with $S \cap \Sigma = \emptyset$, $S_0 \subseteq S$ is the set of *initial states*, $F \subseteq S$ is the set of *final states*, and $\delta : S \times \Sigma \to P(S)$ is the *transition function*.

A non-deterministic finite state automaton M computes a string $x = a_1 \ldots a_n$ similar to its deterministic counterpart. However, M can start in any initial state, and if it happens to enter the state s and reading symbol a, then it can enter any state in $\delta(s, a)$. Therefore, the processing of the input string x can be traced by *all* paths $(s_0, \ldots, s_n)$ in the corresponding transition graph so that $s_0 \in S_0$ and $s_i \in \delta(s_{i-1}, a_i)$ for all $1 \leq i \leq n$.

The multi-step behavior of a non-deterministic finite state automaton M can be formally described by the *extended transition function* $\delta^* : P(S) \times \Sigma^* \to P(S)$ which is inductively defined as follows:

- $\delta^*(S', \varepsilon) = S'$ for each $S' \subseteq S$,
- $\delta^*(S', ax) = \bigcup_{s \in S'} \delta^*(\delta(s, a), x)$ for each $S' \subseteq S$, $a \in \Sigma$, $x \in \Sigma^*$.

In particular, $\delta^*(\{s\}, a) = \delta(s, a)$ for each $s \in S$ and $a \in \Sigma$.

The *language of* M is the set of all strings accepted by M,

$$L(M) = \{x \in \Sigma^* \mid \delta^*(S_0, x) \cap F \neq \emptyset\} . \tag{2.5}$$

Example 2.22. Let $M = (\Sigma, S, \delta, S_0, F)$ be the non-deterministic finite state automaton so that $S = \{s_0, s_1, s_2\}$, $\Sigma = \{a, b\}$, $S_0 = \{s_0, s_1\}$, $F = \{s_2\}$, and transition function δ given by the transition graph in Figure 2.10. The language of M consists of all strings of the form $b^m a w$ and $b^m a b^n a w$, where $m, n \geq 0$ and $w \in \Sigma^*$. ◇

The computing models of deterministic and non-deterministic finite state automata are equivalent, as shown by the following

Theorem 2.23. *Each language accepted by a non-deterministic finite state automaton can also be accepted by a deterministic finite state automaton.*

Proof. Let $M = (\Sigma, S, \delta, S_0, F)$ be a non-deterministic finite state automaton. Define a deterministic finite state automaton $M' = (\Sigma, \mathrm{S}, \delta', s_0', F')$, where

$$\mathrm{S} = P(S) , \tag{2.6}$$

$$\delta'(S', a) = \bigcup_{s \in S'} \delta(s, a) = \delta^*(S', a), \quad S' \in \mathrm{S} , \tag{2.7}$$

$$s_0' = S_0 , \tag{2.8}$$

$$F' = \{S' \subseteq S \mid S' \cap F \neq \emptyset\} . \tag{2.9}$$

Fig. 2.10 Non-deterministic finite state automaton.

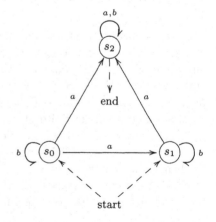

For each $x = a_1 \ldots a_n \in \Sigma^*$, we have $x \in L(M)$ if and only if there is a sequence of subsets $S_1, \ldots, S_n$ of S so that $\delta^*(S_{i-1}, a_i) = S_i$ for all $1 \leq i \leq n$ and $S_n \cap F \neq \emptyset$. The latter is equivalent to $\delta'(S_{i-1}, a_i) = S_i$ for all $1 \leq i \leq n$ and $S_n \in F'$, which in turn is equivalent to $x \in L(M')$. $\square$

2.2.4 Regular Expressions

Regular expressions are formulas which represent regular languages. Let Σ be an alphabet. The set of *regular expressions* over Σ is inductively defined as follows:

- $\emptyset$ is a regular expression
- ε is a regular expression
- For each symbol $a \in \Sigma$, a is a regular expression
- If α and β are regular expressions, then $\alpha\beta$, $\alpha \cup \beta$, and α^* are regular expressions

The *language* of a regular expression γ over Σ is inductively defined as follows:

- If $\gamma = \emptyset$, then $L(\gamma) = \emptyset$
- If $\gamma = \varepsilon$, then $L(\gamma) = \{\varepsilon\}$
- If $\gamma = a$, then $L(\gamma) = \{a\}$
- If $\gamma = \alpha\beta$, then $L(\gamma) = L(\alpha)L(\beta)$
- If $\gamma = \alpha \cup \beta$, then $L(\gamma) = L(\alpha) \cup L(\beta)$
- If $\gamma = \alpha^*$, then $L(\gamma) = L(\alpha)^*$

For each regular expression α, define $\alpha^+ = \alpha\alpha^*$.

Example 2.24. Each finite language can be described by a regular expression. Indeed, if $L = \{x_1, \ldots, x_n\}$ then $\alpha = x_1 \cup \ldots \cup x_n$ is a regular expression so that $L(\alpha) = L$.

The language of the automaton in Example 2.20 is given by the regular expression $\alpha = b^*(ab^*a)^*b^*$, while the language of the machine in Example 2.22 is described by the regular expression $\alpha = b^*a(a \cup b)^* \cup b^*ab^*a(a \cup b)^*$. $\Diamond$

Theorem 2.25. *The set of all languages described by regular expressions equals the set of all languages accepted by finite state automata.*

Proof. Let γ be a regular expression over Σ. Claim that there is a non-deterministic finite state automaton M so that $L(M) = L(\gamma)$. Indeed, this can be proved by induction over the definition of regular expressions. If $\gamma = \emptyset$, $\gamma = \varepsilon$, or $\gamma = a$, $a \in \Sigma$, then $L(\gamma)$ can be described by a finite state automaton with the required property.

Let $\gamma = \alpha\beta$. By induction, there are finite state automata M_1 and M_2 so that $L(M_1) = L(\alpha)$ and $L(M_2) = L(\beta)$. By serializing these automata we obtain a finite state automaton M so that the initial states of M are the initial states of M_1 and the final states of M are the final states of M_2. If $\varepsilon \in L(\alpha)$, then also the initial states of M_2 are initial states of M. Moreover, all states in M_1, which are linked with a final state in M_1, are also connected to the initial states of M_2. Then $L(M) = L(M_1)L(M_2) = L(\alpha)L(\beta) = L(\alpha\beta)$ as required.

Let $\gamma = \alpha \cup \beta$. By induction, there are automata $M_1 = (\Sigma, S_1, \delta_1, s_{01}, F_1)$ and $M_2 = (\Sigma, S_2, \delta_2, s_{02}, F_2)$ with $S_1 \cap S_2 = \emptyset$ so that $L(M_1) = L(\alpha)$ and $L(M_2) = L(\beta)$. Then the automaton $M = (\Sigma, S_1 \cup S_2, \delta_1 \cup \delta_2, \{s_{01}, s_{02}\}, F_1 \cup F_2)$ fulfills $L(M) = L(\alpha \cup \beta)$.

Let $\gamma = \alpha^*$. By induction, there is a finite state automaton M_1 so that $L(M_1) = L(\alpha)$. Define a finite state automaton M so that M has the same initial and final states as M_1. Moreover, each state connected with a final state in M_1 is also linked to the initial states of M_1. If $\varepsilon \notin L(\alpha)$, then add a state which is both initial and final state but not linked with the other states. Then $L(M) = L(M_1)^* = L(\alpha)^* = L(\alpha^*)$.

Conversely, let M be a deterministic finite state automaton over Σ with state set $S = \{s_1, \ldots, s_n\}$ and initial state s_1. Let $1 \leq i, j \leq n$ and $0 \leq k \leq n$. Define L_{ij}^k as the set of all strings $x \in \Sigma^*$ so that $\delta^*(s_i, x) = s_j$ and no intermediate state has an index larger than k (up to s_i and s_j). Claim that each language L_{ij}^k can be described by a regular expression over Σ. Indeed, if $i = j$, then $L_{ii}^0 = \{\varepsilon\} \cup \{a \in \Sigma \mid \delta(s_i, a) = s_i\}$, and if $i \neq j$, $L_{ij}^0 = \{a \in \Sigma \mid \delta(s_i, a) = s_j\}$. Both languages are finite and so can be described by regular expressions. Moreover, we have

$$L_{ij}^{k+1} = L_{ij}^k \cup L_{i,k+1}^k (L_{k+1,k+1}^k)^* L_{k+1,j}^k .$$

Indeed, in order to pass from state s_i to state s_j, the state s_{k+1} is either not used or will be traversed at least once. The latter is described by the language $L_{i,k+1}^k (L_{k+1,k+1}^k)^* L_{k+1,j}^k$.

By induction over k, let α_{ij}^k be a regular expression so that $L(\alpha_{ij}^k) = L_{ij}^k$. Hence, the language L_{ij}^{k+1} can be described by the regular expression

$$\alpha_{ij}^{k+1} = \alpha_{ij}^k \cup \alpha_{i,k+1}^k (\alpha_{k+1,k+1}^k)^* \alpha_{k+1,j}^k .$$

This proves the claim. But the language of M is given by

$$L(M) = \bigcup_{s_i \in F} L_{1,i}^n .$$

Hence, we have

$$L(M) = \bigcup_{s_i \in F} L(\alpha_{1,i}^n) = L\left(\bigcup_{s_i \in F} \alpha_{1,i}^n\right) .$$

$\square$

2.2.5 Stochastic Finite State Automata

Stochastic machines provide a generalization of deterministic machines. Transitions in stochastic machines are based on probability distributions. A *probability distribution* p on a finite set S is a mapping $p : S \to \mathbb{R}_0^+$ so that $\sum_{s \in S} p(s) = 1$.

A *stochastic finite state automaton* is a quintuple $M = (\Sigma, S, P, q_0, q_f)$ so that Σ is an alphabet, S is a finite set of states with $\Sigma \cap S = \emptyset$, q_0 is a *initial probability distribution* on state set S, q_f is a *final probability distribution* on state set S, and P is a *conditional probability distribution* so that $P(\cdot \mid a, s)$ is a probability distribution on the state set S for each pair $(a, s) \in \Sigma \times S$.

Example 2.26. Consider the stochastic finite state automaton M with alphabet $\Sigma = \{a, b\}$, state set $S = \{s_0, s_1\}$, initial and final probability distributions $q_0(s_0) = q_f(s_0) = 1$ and $q_0(s_1) = q_f(s_1) = 0$, and the transition probabilities given in Figure 2.11. $\diamondsuit$

A stochastic finite state automaton M computes a string $x = a_1 \ldots a_n$ similar to its deterministic counterpart. For this, M reads the first symbol starting in the initial state s with probability $q_0(s)$. The computation consists of a series of iterations. In each iteration, if M is in state s with probability q and reads the next input symbol a, then M enters state s' with probability $P(s' \mid a, s) \cdot q$. The computation terminates when the string is exhausted.

The multi-step behavior of a stochastic automaton M can be described by extending the transition probabilities so that for the empty string ε,

$$P(s' \mid \varepsilon, s) = \begin{cases} 1 \text{ if } s = s', \\ 0 \text{ if } s \neq s', \end{cases} \tag{2.10}$$

and for each non-empty string $x = a_1 \ldots a_{n+1}$ over Σ and states $s, s' \in S$,

$$P(s' \mid a_1 \ldots a_n a_{n+1}, s) = \sum_{s'' \in S} P(s' \mid a_{n+1}, s'') P(s'' \mid a_1 \ldots a_n, s) . \tag{2.11}$$

That is, in order to reach state s' from state s by reading the string x, all intermediate states s'' need to be considered that can be reached from s

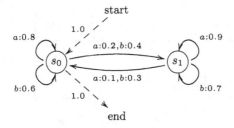

Fig. 2.11 Stochastic finite state automaton.

when the substring $a_1 \ldots a_n$ is read. It follows that $q_f(s')P(s' \mid x, s)q_0(s)$ is the probability that M reaches the final state s' when it reads the input string x starting in the initial state s.

Example 2.27. In view of the stochastic automaton in Example 2.26, the probability of the string ab starting and ending in state s_0 is given by

$$q_f(s_0)P(s_0 \mid ab, s_0)q_0(s_0) =$$
$$= q_f(s_0)\left[P(s_0 \mid b, s_0)P(s_0 \mid a, s_0) + P(s_0 \mid b, s_1)P(s_1 \mid a, s_0)\right]q_0(s_0)$$
$$= 1.0 \cdot [0.6 \cdot 0.8 + 0.3 \cdot 0.2] \cdot 1.0 = 0.54 \;.$$

$\diamondsuit$

A stochastic finite state machine M accepts an input string x with the probability

$$P(x) = \sum_{s,s' \in S} q_f(s')P(s' \mid x, s)q_0(s) \;. \tag{2.12}$$

This is the probability that M enters a final state when it reads the string x starting from an initial state.

A stochastic finite state automaton M becomes a *stochastic finite parser* by taking into account a *threshold value* $\lambda \geq 0$. A stochastic finite parser M *accepts* a string x if the probability $P(x)$ of the string being accepted

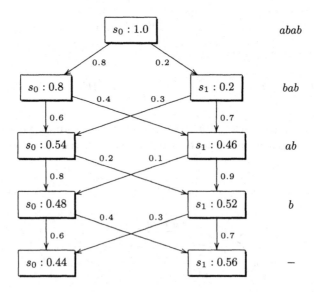

Fig. 2.12 Computation of the string *abab* in the stochastic finite automaton M (Fig. 2.11). This string is accepted with probability 0.44.

exceeds the threshold value. The *language* of M with given threshold value λ is defined by all accepted strings,

$$L_\lambda(M) = \{x \in \Sigma^* \mid P(x) > \lambda\}\,.$$

Example 2.28. Let $\lambda \geq 0$. The language of the parser in Example 2.26 with threshold value λ consists of all strings x over Σ so that $P(x) = P(s_0 \mid x, s_0) > \lambda$, as s_0 is with certainty the initial and final state. The probability of the string *abab* is calculated in Figure 2.12.   ◇

2.3 Computability

This section introduces some venerable formalisms for computing that reflect what any physical computing device is capable of doing. This includes the construction of a universal machine that provides a programmable computer.

2.3.1 Turing Machines

Turing machines were invented by A. Turing (1936) as a thought experiment about the limits of mechanical computation. Studying their abstract properties yields many insights into computational theory. A Turing machine consists of a *tape* which is divided into cells. Each *cell* contains a symbol from a common alphabet. The alphabet contains a special *blank* symbol written as "□". The tape is arbitrarily long so that the machine always has as much tape as possible. A *head* can read and write symbols on the tape and move the tape to the left or to the right one cell at a time. A *state register* contains the machine's transition function (Fig. 2.13).

A *Turing machine* is a 7-tuple $M = (\Sigma, S, \Gamma, \delta, s_0, \square, F)$ so that Σ is an alphabet, S is a finite set of *states*, $\Gamma \supseteq \Sigma$ is the *tape alphabet*, $s_0 \in S$ is the *initial state*, $\square \in \Gamma \setminus \Sigma$ is the *blank*, $F \subseteq S$ is the set of *final states*, and

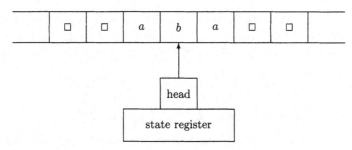

Fig. 2.13 A Turing machine.

$\delta : S \times \Gamma \to S \times \Gamma \times \{L, R, N\}$ is the *transition function*. This function may be partially defined, but this does not alter what a Turing machine can do.

The equation $\delta(s, a) = (s', b, r)$ is referred to as a *transition rule* and can be interpreted as follows: If M is in state s and reads the symbol a from the band tape, then M enters the state s', writes the symbol b on the band tape, and moves the head according to the value of r: left (L), right (R), or non-moving (N).

The overall state of a Turing machine at any stage in a computation is described by a *configuration*, which is given as a string in $\Gamma^* S \Gamma^*$. The configuration $\alpha s \beta$ means that $\alpha \beta$ is the contents of the band tape, s is the state of M, and the head is located on the first symbol of β. Each computation of a Turing machine is started in an *initial configuration* $s_0 x$, where x is an input string over Σ.

The one-step behavior of a Turing machine M is formally described by a binary relation on the set of configurations of M as follows:

$$a_1 \ldots a_m s b_1 \ldots b_n \vdash_M \begin{cases} a_1 \ldots a_m s' c b_2 \ldots b_n & \text{if } \delta(s, b_1) = (s', c, N), \\ & m \geq 0, n \geq 1, \\ a_1 \ldots a_m c s' b_2 \ldots b_n & \text{if } \delta(s, b_1) = (s', c, R), \\ & m \geq 0, n \geq 2, \\ a_1 \ldots a_{m-1} s' a_m c b_2 \ldots b_n & \text{if } \delta(s, b_1) = (s', c, L), \\ & m \geq 1, n \geq 1 . \end{cases}$$

Moreover, if $n = 1$ and M moves to the right, then M meets a blank,

$$a_1 \ldots a_m s b_1 \vdash_M a_1 \ldots a_m c s' \square, \quad \text{if } \delta(s, b_1) = (s', c, R) . \tag{2.13}$$

If $m = 0$ and M moves to the left, then M also meets a blank,

$$s b_1 \ldots b_n \vdash_M s' \square c b_2 \ldots b_n, \quad \text{if } \delta(s, b_1) = (s', c, L) . \tag{2.14}$$

Furthermore, if M enters a configuration in which no transition rule is applicable, then M halts. The multi-step behavior of M is given by the reflexive, transitive closure $\vdash_M^*$ of $\vdash_M$. A *computation* of M is a finite sequence of configurations $(\gamma_0, \ldots, \gamma_n)$ so that γ_0 is an initial configuration and $\gamma_i \vdash_M \gamma_{i+1}$ for each $0 \leq i \leq n - 1$.

Example 2.29. Consider the Turing machine M with input alphabet $\Sigma = \{0, 1\}$, state set $S = \{s_1, s_2, s_3\}$, tape alphabet $\Gamma = \{0, 1, \square\}$, initial state s_1, final state set $F = \{s_2\}$, and partially defined transition function δ given as follows:

$$\delta(s_1, 0) = (s_1, 0, R), \quad \delta(s_3, 0) = (s_3, 1, L),$$
$$\delta(s_1, 1) = (s_1, 1, R), \quad \delta(s_3, 1) = (s_3, 0, L),$$
$$\delta(s_1, \square) = (s_3, \square, L), \quad \delta(s_3, \square), = (s_2, \square, R) .$$

The machine M reverses each input string $x \in \{0, 1\}^*$. For instance, given the input $x = 101$, the machine produces the string 110 by the following

computation: $s_1 101 \vdash_M 1s_1 01 \vdash_M 10s_1 1 \vdash_M 101s_1 \square \vdash_M 10s_3 1 \vdash_M 1s_3 00 \vdash_M$
$s_3 110 \vdash_M s_3 \square 010 \vdash_M s_2 010.$ ◇

The *language* of a Turing machine M is defined by all input strings whose computation ends in a final state,

$$L(M) = \{x \in \Sigma^* \mid s_0 x \vdash_M^* \alpha s \beta, \alpha, \beta \in \Gamma^*, s \in F\} . \tag{2.15}$$

Example 2.30. In view of the previous example, the language of the Turing machine M equals $\{0, 1\}^*$.   ◇

2.3.2 Universal Turing Machines

Alan Turing was the first to show that universal Turing machines exist. A Turing machine is termed *universal* if it is able to simulate any Turing machine. This was a remarkable finding at that time (1936) and led to the notion of programmable computer.

First, notice that each Turing machine can be thought of as having a finite number of tapes (Fig. 2.14). Using multiple tapes does not extend what a Turing machine can do.

Second, notice that each Turing machine can be encoded by a binary string. For this, let M be a Turing machine with binary input alphabet $\Sigma = \{0, 1\}$. Assume that the states are $s_1, \ldots, s_r$ for some r, where s_1 is the initial state and s_2 is the unique final state, as we assume that the machine halts if it enters an accepting state. Moreover, let the tape symbols be $z_1, \ldots, z_s$ for some s, with $z_1 = 0$, $z_2 = 1$, and $z_3 = \square$, and refer to the direction L as D_1, R as D_2, and N as D_3. Therefore, each transition rule has the form

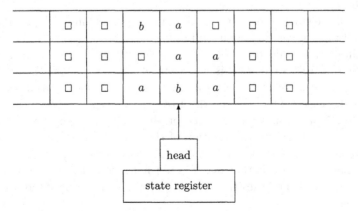

Fig. 2.14 A Turing machine with three tapes.

$\delta(s_i, z_j) = (s_k, z_l, D_m)$ for some integers i, j, k, l, and m. This rule can be encoded by the string $0^i 10^j 10^k 10^l 10^m$. Since all integers i, j, k, l, and m are positive, the encoded string never contains two or more consecutive 1's. Finally, the entire Turing machine M can be encoded by concatenating the transition rules and separating them by pairs of 1's:

$$t_1 11 t_2 11 \ldots 11 t_{n-1} 11 t_n \, , \tag{2.16}$$

where each $t_1, \ldots, t_n$ are the encodings of the transitions in M.

Example 2.31. The rules of the Turing machine M in Example 2.29 are encoded as follows:

0101010100,   00010100010010,

010010100100,   00010010001010,

010001000100010,   00010001001000100 .

Theorem 2.32. *Universal Turing machines exist.*

Proof. Define the *universal language* L_u as the set of all binary strings that encode pairs (M, x), where M is a Turing machine with binary input alphabet and x is a binary input string so that x lies in $L(M)$. Claim that there is a Turing machine U so that $L(U) = L_u$. Indeed, assume that U has multiple tapes. More precisely, the first tape initially holds the transitions of M, along with the string x. The second tape stores the simulated tape of M, and the third tape holds the state of M. The operations of U can be summarized as follows:

1. Examine the input to check whether the encoding of M is legitimate. If not, U halts without acceptance.
2. Initialize the second tape to contain the input string x in its encoded form (i.e., for each 0 in x place 10 on the tape and for each 1 in x place 100 there).
3. Place 0, the start state of M, on the third tape, and move the head of U's second tape to the first simulated cell.
4. To simulate a transition of M, U searches on its first tape for a string $0^i 10^j 10^k 10^l 10^m$ so that 0^i is the state of the third tape, and 0^j is the tape symbol of M that begins at the position of the second tape. If so, U changes the contents of the third tape to 0^k, replaces 0^j on the second tape by 0^l, and keeps the head (N) on the second tape or moves the head on the second tape to the position of the next 1 to the left (L) or to the right (R).
5. If M has no transition that matches the simulated state and tape symbol, then in step 4, no transition will be found. Thus, M halts and U does likewise.
6. If M enters its accepting state, then U accepts (M, x).

In this way, U simulates M on x so that U accepts the encoded pair (M, x) if and only if M accepts x. This proves the claim.   □

Example 2.33. A Turing machine with a proven universal computation and the smallest number of states times number of symbols is a 2-state, 5-symbol Turing machine discovered by S. Wolfram (Fig. 2.15).   ◇

2.3.3 Church's Thesis

Computability theory mainly addresses the computation of (arithmetic) functions. A function $f : \mathbb{N}_0^k \to \mathbb{N}_0$ is called *computable* if there is an algorithm, say in the form of a C program, which computes the function f (i.e., the algorithm starts with a k-tuple $(n_1, \ldots, n_k) \in \mathbb{N}_0^k$ as input and terminates after a finite number of steps providing the output $f(n_1, \ldots, n_k)$). Such a function can be partially defined, that is, provide no value for some input. If the corresponding algorithm starts with a k-tuple that does not belong to the domain of the function, then the algorithm is assumed to enter an infinite loop.

Example 2.34. The following function corresponding to the Archimedes constant π is computable,

$$f_\pi(m) = \begin{cases} 1 \text{ if } m \text{ is a prefix of the decimal evaluation of } \pi, \\ 0 \text{ otherwise.} \end{cases} \tag{2.17}$$

For instance, $f_\pi(314) = 1$ and $f_\pi(2) = 0$. Indeed, there are approximation methods for computing the first m positions of the constant π.   ◇

A function $f : \mathbb{N}_0^k \to \mathbb{N}_0$ is called *Turing computable* if there is a Turing machine M so that for all $n_1, \ldots, n_k \in \mathbb{N}_0$, we have $f(n_1, \ldots, n_k) = m$ if and only if

$$s_0 \mathrm{bin}(n_1)\# \ldots \#\mathrm{bin}(n_k) \vdash_M^* s\,\mathrm{bin}(m), \quad s \in F, \tag{2.18}$$

where $\mathrm{bin}(n)$ denotes the binary representation of $n \in \mathbb{N}_0$ and $\#$ is an auxiliary symbol. Notice that if the function f is partially defined, then for each

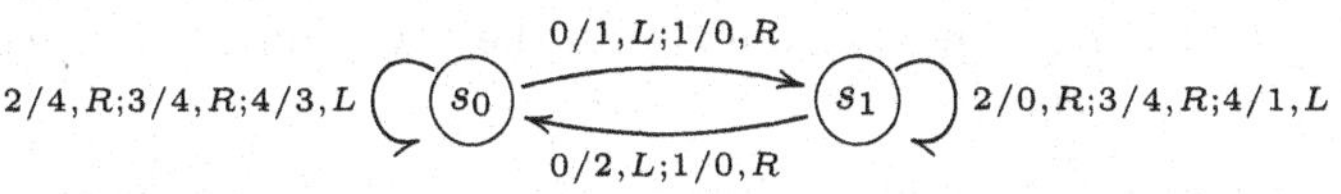

Fig. 2.15 Universal Turing machine.

k-tuple $(n_1, \ldots, n_k)$ not lying in the domain of f, the Turing machine runs into an infinite loop.

Example 2.35. The complement function $n \mapsto \bar{n}$ complementing each bit in the binary representation of a non-negative integer n is Turing computable as demonstrated by the Turing machine in Example 2.29.

The nowhere defined function is Turing computable by taking the Turing machine M whose transitions are of the form $\delta(s_0, a) = (s_0, a, R)$, $a \in \Gamma$. The machine M moves to the right without stopping. $\diamond$

The thesis stated by A. Church in the 1940s hypothesizes that any (intuitively) computable function is Turing computable, and vice versa.

However, not every function is computable per se. To see this, consider a real number r, $0 \le r \le 1$, and define the function

$$f_r(m) = \begin{cases} 1 \text{ if } m \text{ is a prefix of the decimal evaluation of } r, \\ 0 \text{ otherwise.} \end{cases} \quad (2.19)$$

On the one hand, the set of Turing programs is denumerable. A set A is called *denumerable* if there is a bijective mapping $g : \mathbb{N} \to A$.

Theorem 2.36. *The set of all strings over a finite alphabet is denumerable.*

Proof. Let Σ be a finite set. The sets Σ^n, $n \ge 0$, are finite and form a partition of Σ^*. Let $a_{n1}, a_{n2} \ldots, a_{nm_n}$ be the elements of Σ^n. The mapping $g : \Sigma^* \to \mathbb{N}_0$, defined as $a_{nj} \mapsto m_0 + \ldots + m_{n-1} + j$, is bijective. $\square$

On the other hand, the real-valued interval $[0, 1]$ is *non-denumerable*. To see this, we use a diagonalization argument first proposed by G. Cantor, the founder of modern set theory, in the 19th century.

Theorem 2.37. *The closed interval $[0, 1]$ is non-denumerable.*

Proof. Suppose there is a bijection $g : \mathbb{N} \to [0, 1]$, which assigns to each positive integer n a real number $g(n) = r^{(n)} \in [0, 1]$. Each number $r^{(n)}$ can be represented as

$$r^{(n)} = 0.r_1^{(n)} r_2^{(n)} r_3^{(n)} \ldots, \quad 0 \le r_i^{(n)} \le 9 .$$

Define the real number $x = 0.x_1 x_2 x_3 \ldots$ as follows:

$$x_n = \begin{cases} 1 \text{ if } r_n^{(n)} \text{ is even,} \\ 0 \text{ otherwise.} \end{cases}$$

The number x lies in $[0, 1]$ but does not appear in the list $r^{(0)}, r^{(1)}, r^{(2)}, \ldots$, because by definition, x differs from $r^{(n)}$ in the nth position $r_n^{(n)}$. Therefore, g is not bijective, a contradiction. $\square$

It follows that there exist real numbers r so that f_r is not Turing computable. By Church's thesis, there exist real numbers r so that f_r is not computable per se.

2.3.4 Register Machines

Register machines provide an abstract machine model whose computational power is equivalent to that of Turing machines. A *register machine* consists of an infinite number of registers $r_0, r_1, r_2, \ldots$ each of which holds a single, non-negative integer. A separate state register stores the current instruction to be executed. A *computation* of a register machine consists of a finite sequence of instructions. Historically, the most common *instruction set* is the following:

- inc(i): Increment the value of register r_i by 1
- dec(i, m): If the value of register r_i is greater than 0, then decrement this value by 1. Otherwise, go to the mth instruction
- jump(m): Go to the mth instruction
- halt: Halt the computation

Example 2.38. Consider the natural-valued successor function $n \mapsto n + 1$. Suppose the input is stored in register r_1 and the output stored in register r_0. A register machine program computing the successor function is specified by the algorithm SUCCESSORREGMACHINE. $\diamondsuit$

Algorithm 2.1 SUCCESSORREGMACHINE
___
Input: non-negative integer n, stored in register r_1
  1: dec($1, 4$)
  2: inc(0)
  3: jump(1)
  4: inc(0)
  5: halt
  6: **return** r_0

2.3.5 Cellular Automata

A cellular automaton is another abstract computational model whose computational power is equivalent to that of Turing machines. This model was introduced by S. Ulam and J. von Neumann in the 1940s. A *cellular automaton* consists of an infinite regular grid of cells. The grid is embedded into a finite-dimensional space, and each cell is in one of a finite number of states. All cells perform the same transition function on the states of neighboring cells. These neighbors are selected relative to the location of the cell and do not change during computation. A *configuration* of a cellular automaton is the collection of the states of all cells. A cellular automaton evolves in a discrete manner so that all cells provide a transition at time t leading from the configuration at time t to a new configuration at time $t + 1$.

The *Game of Life* is a two-dimensional grid-like cellular automaton introduced by J. Conway in the 1970s. Each cell is in one of two states, live or dead, and the neighborhood of a cell is constituted by the eight surrounding cells. Any live cell dies if it has fewer than two live neighbors (loneliness) or more than three live neighbors (overcrowding). Any live cell lives unchanged if it exhibits two or three live neighbors. Any dead cell comes to life if it has three neighbors. The Game of Life exhibits *gliders*, which are arrangements of cells that move themselves across the grid. Gliders can interact so that they perform computations. It can be shown that the Game of Life can emulate a universal Turing machine.

An *elementary cellular automaton* is a one-dimensional cellular automaton so that the state set is binary and the neighborhood of a cell is formed by the cell itself and its left and right neighbors. Thus, there are $2^3 = 8$ binary patterns for the neighborhood and hence $2^8 = 256$ transition functions. Each of these 256 elementary cellular automata is indexed with an 8-bit binary number, which describes the transition function, with the eight states listed in reverse counting order.

Example 2.39. The 30th elementary cellular automaton (30=00011110) is given by the transition function in Table 2.1. For instance, the pattern 110 describes the state of a cell and its left and right neighbors (i.e., the left and center cells are on (1), and the right cell is off (0)). The transition rule says that the middle cell will be off at the next step (Fig. 2.16).   ◇

A *one-dimensional blocked cellular automaton* (BCA) C is based on an infinite linear array of cells and a finite set of transition rules $(u, v) \rightarrow_C (x, y)$ over an alphabet Σ. Let $c_t(x)$ denote the symbol at cell x in configuration c_t at time t. A BCA uses the transition rules to rewrite pairs of cells in c_t, alternating between even and odd alignments of the pairings (Fig. 2.17): For even t and even x, and for odd t and odd x,

$$(c_t(x), c_t(x+1)) \rightarrow_C (c_{t+1}(x), c_{t+1}(x+1)) . \tag{2.20}$$

The input of a BCA computation is a configuration which assigns the blank symbol to all, but a finite number of cells.

Theorem 2.40. *Each Turing machine can be simulated by a one-dimensional blocked cellular automaton.*

Proof. Define a one-dimensional BCA that reproduces the space-time history of the tape of a Turing machine.   □

Table 2.1 Transition function of the 30th elementary cellular automaton.

Current states	111	110	101	100	011	010	001	000
New state for center	0	0	0	1	1	1	1	0

1						1	1	1				1			1	1	8
	1	1	1	1	1	1			1	1	1	1		1	1		7
		1			1					1			1	1			6
			1	1	1		1	1	1	1		1	1				5
				1				1			1	1					4
					1	1	1	1		1	1						3
						1			1	1							2
							1	1	1								1
								1									0

Fig. 2.16 Evolution of the 30th elementary cellular automaton starting from a single on-cell after nine steps.

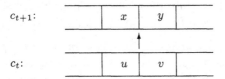

Fig. 2.17 Transition rule $(u, v) \to_C (x, y)$ in BCA.

2.4 Formal Grammars

Formal grammars were first studied by the linguist N. Chomsky in the late 1950s and provide another formalism for universal computation.

2.4.1 Grammars and Languages

A *(formal) grammar* is a quadruple $G = (\Sigma, V, P, S)$ consisting of a finite set of *terminal symbols* Σ, a finite set of *non-terminal symbols* V with $\Sigma \cap V = \emptyset$, a finite set of *production rules* $P \subseteq (\Sigma \cup V)^+ \times (\Sigma \cup V)^*$, and a *start symbol* $S \in V$. A production rule $(u, v) \in P$ is also written as $u \to_G v$.

A grammar allows the derivation of strings over Σ from the start symbol. For this, let $u, v \in (\Sigma \cup V)^*$. We write $u \Rightarrow_G v$, if u and v have the form $u = xyz$ and $v = xy'z$, where $x, z \in (\Sigma \cup V)^*$ and $y \to_G y'$.

A *derivation* of a string $w \in (\Sigma \cup V)^*$ in G is a finite sequence of strings $(w_0, \ldots, w_n)$ so that $w_0 = S$, $w_n = w$, and $w_i \Rightarrow_G w_{i+1}$ for each $0 \leq i \leq n-1$. Let $\Rightarrow_G^*$ denote the reflexive, transitive closure of $\Rightarrow_G$. Then a derivation of w in G exists if and only if $S \Rightarrow_G^* w$. The *language* of G is the set of all strings over Σ derived in G,

$$L(G) = \{w \in \Sigma^* \mid S \Rightarrow_G^* w\} . \tag{2.21}$$

Example 2.41. Consider the grammar $G = (\Sigma, V, P, S)$ with $\Sigma = \{a, b, c\}$, $V = \{S, B\}$, and $P = \{(S, aBSc), (S, abc), (Ba, aB), (Bb, bb)\}$. Some derivations in G are

- $S \Rightarrow_G abc$,
- $S \Rightarrow_G aBSc \Rightarrow_G aBabcc \Rightarrow_G aaBbcc \Rightarrow_G aabbcc$,
- $S \Rightarrow_G aBSc \Rightarrow_G aBaBScc \Rightarrow_G aBaBabccc \Rightarrow_G^* aaaBBbccc \Rightarrow_G^*$ $aaabbbccc$.

The language of G consists of all strings of the form $a^n b^n c^n$, $n \geq 1$. $\Diamond$

2.4.2 Chomsky's Hierarchy

The Chomsky hierarchy introduced by N. Chomsky (1956) is a containment hierarchy of four classes of formal grammars given as follows:

- Each grammar is automatically of *type-0* or *recursively enumerable*.
- A grammar G is of *type-1* or *context-sensitive* if each production rule $u \rightarrow_G v$ is *length preserving*, that is, $|u| \leq |v|$, where $u, v \in (\Sigma \cup V)^*$.
- A grammar G is of *type-2* or *context-free* if each production rule has the form $A \rightarrow_G w$, where $A \in V$ and $w \in (\Sigma \cup V)^*$.
- A grammar G is of *type-3* or *regular* if each production rule has the shape $A \rightarrow_G a$ or $A \rightarrow_G aB$, where $A, B \in V$ and $a \in \Sigma \cup \{\varepsilon\}$.

A language L is of type-0 (type-1, type-2, type-3) if there is a respective grammar G of type-0 (type-1, type-2, type-3) so that $L(G) = L$.

Example 2.42. The language $L = \{a^n \mid n \geq 0\}$ is regular, because L is generated by the grammar $G = \{\{a\}, \{S\}, \{(S, aS), (S, \varepsilon)\}, S)$. $\Diamond$

Example 2.43. The language $L = \{a^n b^n \mid n \geq 1\}$ is context-free, as L is generated by the grammar $G = \{\{a, b\}, \{S\}, \{(S, aSb), (S, ab)\}, S)$. $\Diamond$

Example 2.44. The language $L = \{a^n b^n c^n \mid n \geq 1\}$ is context-sensitive, for L is generated by the grammar in Example 2.41. $\Diamond$

The following inclusions are obvious.

Theorem 2.45. *Regular languages are context-free, context-free languages are context-sensitive, and context-sensitive languages are recursively enumerable.*

A context-sensitive grammar is in *Kuroda normal form* if all production rules are of the form $A \rightarrow_G a$, $A \rightarrow_G B$, $A \rightarrow_G BC$, or $AB \rightarrow_G CD$, where a is a terminal symbol and A, B, C, and D are non-terminal symbols.

Theorem 2.46. *Each grammar in Kuroda normal form is context-sensitive. For each context-sensitive grammar G with $\varepsilon \notin L(G)$ there exists a grammar G' in Kuroda normal form so that $L(G') = L(G)$.*

Proof. The first statement follows from the definitions. Let $G = (\Sigma, V, P, S)$ be a context-sensitive grammar with $\varepsilon \notin L(G)$. Define a grammar $G' = (\Sigma, V', P', S)$ as follows:

- Each terminal symbol a in P is replaced by a new non-terminal symbol A_a, and the production rule $A_a \rightarrow_{G'} a$ is added.
- Each production rule of the form $A \rightarrow_G A_1 \ldots A_k$, $k \geq 3$, is replaced by the following production rules: $A \rightarrow_{G'} A_1 B_1$, $B_i \rightarrow_{G'} A_{i+1} B_{i+1}$ for $1 \leq i \leq k-2$, and $B_{k-1} \rightarrow_{G'} A_k$, where $B_1, \ldots, B_{k-1}$ are new non-terminals.
- Each production rule of the shape $A_1 \ldots A_k \rightarrow_G B_1 \ldots B_l$, $2 \leq k \leq l$ and $l \geq 3$, is substituted by the following production rules: $A_1 A_2 \rightarrow_{G'} B_1 C_1$, $C_i A_{i+2} \rightarrow_{G'} B_{i+1} C_{i+1}$ for $1 \leq i \leq k-2$, $C_i \rightarrow_{G'} B_{i+1} C_{i+1}$ for $k-1 \leq i \leq l-2$, and $C_{l-1} \rightarrow_{G'} B_l$, where $C_1, \ldots, C_{k-2}$ are new non-terminals.

The grammar G' has the required properties. $\square$

2.4.3 Grammars and Machines

Each class of type-i grammars corresponds to an abstract machine model (Table 2.2). While automata are analytic in the sense that they describe how to *read* a language, grammars are synthetic in the sense that they address how to *write* a language.

Theorem 2.47. *The languages described by regular grammars can be recognized by finite state automata, and vice versa.*

Proof. Let $M = (\Sigma, S, \delta, s_0, F)$ be a finite state automaton. Define a regular grammar $G = (\Sigma, V, P, S)$ with $V = S$, $S = s_0$, and production set $P = \{(s, as') \mid s \xrightarrow{a}_M s'\} \cup \{(s, a) \mid s \xrightarrow{a}_M s', s' \in F\}$. A string $x = a_1 \ldots a_n \in \Sigma^*$ lies in $L(M)$ if and only if there is a sequence of states $(s_0, \ldots, s_n)$ so that $\delta(s_{i-1}, a_i) = s_i$, $1 \leq i \leq n$, and $s_n \in F$. This means that there is a derivation in G so that $s_0 \Rightarrow_G a_1 s_1 \Rightarrow_G \ldots \Rightarrow_G a_1 \ldots a_{n-1} s_{n-1} \Rightarrow x$, which is equivalent to $x \in L(G)$.

Conversely, let $G = (\Sigma, V, P, S)$ be a regular grammar. Define a non-deterministic finite state automaton $M = (\Sigma, S, \delta, S_0, F)$ so that $S = V \cup \{X\}$,

Table 2.2 The Chomsky hierarchy.

Grammar	Language	Machine
type-0	recursively enumerable	Turing machine
type-1	context-sensitive	linearly restricted Turing machine
type-2	context-free	pushdown automaton
type-3	regular	finite state automaton

$X \notin V$, $S_0 = \{S\}$, $F = \{S, X\}$ if $(S, \varepsilon) \in P$ and $F = \{X\}$ if $(S, \varepsilon) \notin P$, and $\delta(A, a) = \{B \mid (A, aB) \in P\} \cup \{X \mid (A, a) \in P\}$. A non-empty string $x = a_1 \ldots a_n \in \Sigma^*$ belongs to $L(G)$ if and only if there is a derivation in G of the form $S \Rightarrow_G a_1 A_1 \Rightarrow_G \ldots \Rightarrow_G a_1 \ldots a_{n-1} A_{n-1} \Rightarrow x$. This means that there is a sequence of states $(A_1, \ldots, A_{n-1})$ so that $A_1 \in \delta(S, a_1)$, $A_i \in \delta(A_{i-1}, a_i)$, $2 \leq i \leq n-1$, and $X \in \delta(A_{n-1}, a_n)$, which is equivalent to $x \in L(M)$. Moreover, $\varepsilon \in L(G)$ if and only if $\varepsilon \in L(M)$. □

Theorem 2.48. *Formal grammars generate recursively enumerable languages and can be recognized by Turing machines, and vice versa.*

Proof. Let G be a type-0 grammar. We informally describe a Turing machine M, which accepts $L = L(G)$. Let x be an input string. The Turing machine M iteratively picks (in a deterministic fashion) a rule $(u, v) \in P$ and searches the tape if it contains v as a substring. If so, M replaces the string v with the string u. The machine terminates exactly when the tape only contains the start symbol S. But $x \in L$ if and only if there is a derivation $S \Rightarrow_G^* x$. By reversing the order of this derivation, we obtain a computation of M so that $x \in L(M)$. □

Theorem 2.49 (Type-3 Pumping Lemma). *If L is a regular language over Σ, then there is a non-negative integer n so that each string $x \in L$ of length at least n can be written in the form $x = uvw$ so that $|v| \geq 1$, $|uv| \leq n$, and $uv^i w \in L$ for all integers $i \geq 0$.*

Proof. Let L be a regular language. By Theorem 2.47, there is a finite state automaton M so that $L = L(M)$. Let n be the number of states of M. Consider a string x of length n accepted by M, so when reading x, the automaton runs through $n + 1$ states. By the pigeonhole principle, at least two of these states are the same, which shows that M runs in a loop. Consider the decomposition $x = uvw$ so that the state after reading u equals the state after reading uv. This decomposition has the required properties. □

Example 2.50. The language $\{a^n b^n \mid n \geq 1\}$ is context-free in Example 2.43, but not regular in view of the Pumping Lemma 2.49. ◇

Theorem 2.51 (Type-2 Pumping Lemma). *If L is a context-free language over Σ, then there is a non-negative integer n so that each string $x \in L$ of length at least n can be written in the form $x = uvwyz$ so that $|vy| \geq 1$, $|uwy| \leq n$, and $uv^i wy^i z \in L$ for all integers $i \geq 0$.*

Example 2.52. The language $\{a^n b^n c^n \mid n \geq 1\}$ is context-sensitive in Example 2.44, but not context-free in view of the Pumping Lemma 2.51. ◇

2.4.4 Undecidability

In view of the binary encoding of Turing machines in the section on universal Turing machines, define the *ith Turing machine* as the Turing machine M_i

whose encoding is the ith binary string w_i. However, many non-negative integers do not correspond to the encoding of any Turing machine such as 11101, because it does not begin with 0. If the ith binary string is not a valid encoding of a Turing machine, then the ith Turing machine is taken to be the Turing machine with one state and no transitions, and thus $L(M_i) = \emptyset$. Define the *diagonalization language* as the language L_d consisting of all binary strings w_i so that w_i is not in $L(M_i)$.

Theorem 2.53. *The diagonalization language L_d is not recursively enumerable.*

Proof. Apply a diagonalization argument. Assume that L_d is recursively enumerable. Then by Theorem 2.48, there is a Turing machine M so that $L(M) = L_d$. Consequently, $M = M_i$ for some non-negative integer i. But by definition $w_i \in L_d$ if and only if $w_i \notin L(M)$ contradicting the hypothesis. $\square$

A language L is called *recursive* if there is a Turing machine M so that $L = L(M)$ and for any input string x that lies in L the machine accepts and halts, while for any input string x that lies not in L the machine accepts and never enters a final state.

Theorem 2.54. *Each recursive language is recursively enumerable. Each regular, context-free, or context-sensitive language is recursive.*

Theorem 2.55. *If L is a recursive language, so is the complement $\overline{L}$.*

Proof. Let M be a Turing machine so that $L(M) = L$. Derive a Turing machine $\overline{M}$ from M by swapping the final and non-final states. Since M is guaranteed to halt, $\overline{M}$ is also guaranteed to halt. Furthermore, $\overline{M}$ accepts those strings that M does not accept, and vice versa. $\square$

Theorem 2.56. *The universal language L_u is recursively enumerable but not recursive.*

Proof. In view of the proof of Theorem 2.32, the language L_u is recursively enumerable. Assume that L_u is recursive. Then by Theorem 2.55, the complement $\overline{L}_u$ is recursive, too. Thus there exists a Turing machine M so that $L(M) = \overline{L}_u$. For the input w_i, the machine M determines whether M_i accepts w_i. That is, M accepts w_i if and only if $w_i \notin L_u$. Thus M accepts the diagonalization language L_d contradicting Theorem 2.53. $\square$

Recursive languages capture the notion of decidability. A language L is called *decidable* if it is a recursive language. Indeed, a Turing machine accepting a recursive language L computes the *characteristic function* of L,

$$\chi_L(x) = \begin{cases} 1 \text{ if } x \in L, \\ 0 \text{ otherwise.} \end{cases} \tag{2.22}$$

A language is called *undecidable* if it is not recursive. A recursively enumerable language L is *semi-decidable* in the sense that a Turing machine accepting L will halt if the input lies in L and will otherwise eventually run forever. For instance, the universal language L_u is semi-decidable by Theorem 2.56, while the diagonalization language L_d is not semi-decidable by Theorem 2.53. The most prominent undecidable problem is the so-called halting problem. For this, consider the binary language $L_h = \{w_i \mid M_i \text{ with input } w_i \text{ halts}\}$.

Theorem 2.57 (Halting Problem). *The language L_h is undecidable.*

Proof. Suppose $L = L_h$ is recursive. Then there is a Turing machine M computing χ_L. Define a Turing machine M' with binary input so that M' with input x halts if $\chi_L(x) = 0$, and M' with input x never halts if $\chi_L(x) = 1$. Apply a diagonalization argument. For this, there is a non-negative integer i so that $M' = M_i$. Then M' with input w_i halts if and only if $\chi_L(w_i) = 0$. That is, $w_i \notin L$, which is equivalent to M_i with input w_i never halts. A contradiction. □

A concrete undecidable problem that has nothing to do with the abstraction of the Turing machine is *Post's correspondence problem* (PCP) introduced by E. Post in 1946. Let Σ be an alphabet. An instance of PCP consists of two sequences of strings over Σ, $U = (u_1, \ldots, u_n)$ and $V = (v_1, \ldots, v_n)$. A *solution* of this instance of PCP is a sequence of one or more integers $(i_1, i_2, \ldots, i_k)$ so that $u_{i_1} u_{i_2} \ldots u_{i_k} = v_{i_1} v_{i_2} \ldots v_{i_k}$.

Example 2.58. Consider an instance of the binary PCP given by the sequences $U = (1, 10, 10111)$ and $V = (111, 0, 10)$. This instance has the solution $(3, 1, 1, 2)$, since $u_3 u_1 u_1 u_2 = 101111110 = v_3 v_1 v_1 v_2$.   ◇

PCP can be considered as a language. To this end, the alphabet of a PCP instance with up to 2^m symbols can be encoded by m-ary binary numbers. In this way, each PCP instance can be encoded by a string in a 3-symbol alphabet consisting of 0, 1, and the "comma" symbol. Such a string starts with the number of pairs, n, in binary, followed by a comma, followed by both sequences of strings so that the strings are separated by commas and their symbols are encoded as binary numbers. For instance, the above binary PCP is encoded by the string $11, 1, 10, 10111, 111, 0, 10$.

Theorem 2.59. *Post's correspondence problem is undecidable.*

Proof. (Sketch) First, PCP is equivalent to a *modified Post's correspondence problem* (MPCP). An instance of MPCP consists of two sequences of strings over Σ, $U = (u_1, \ldots, u_n)$ and $V = (v_1, \ldots, v_n)$, and a solution of this instance is a sequence of integers $(i_1, i_2, \ldots, i_k)$, $k \geq 0$, so that $u_1 u_{i_1} u_{i_2} \ldots u_{i_k} = v_1 v_{i_1} v_{i_2} \ldots v_{i_k}$. Thus the strings u_1 and v_1 are forced to be prefixes of each solution.

Let M be a Turing machine so that $L(M) = L_u$. Claim that there is an MPCP instance so that M accepts a string x if and only if the MPCP instance has a solution. Indeed, the computation of the Turing machine M on input x can be simulated by an MPCP instance, where the forced strings u_1 and v_1 provide the initial state: $u_1 = \#$ and $v_1 = \#s_0x\#$. As L_u is undecidable, the result follows. $\square$

Another prominent undecidable problem is *Wang's tiling problem*, first stated by H. Wang in the 1960s. A tiling of the plane is a collection of plane figures that fills the plane with no gaps and overlaps. Today, tilings are perhaps best known for their use in the art of the Dutch painter M.C. Escher. The tiling problem considers plane figures given by unit-sized squares, called *tiles*, with colored edges. The task is to aggregate tiles by placing them next to each other without space so that neighboring tiles have the same color along their common edges. These aggregates are termed *tilings*. Notice that the tiles cannot be rotated or reflected, and the set of different tiles in a tiling is assumed to be finite. Wang's tiling problem is to decide whether a tiling of the plane can be obtained from a given finite set of tiles (Fig. 2.18).

Theorem 2.60. *Wang's tiling problem is undecidable.*

Proof. Create a set of tiles that fit together uniquely to reproduce the space-time history of the tape of a Turing machine. Consequently, if the Turing machine halts then the attemped tiling gets stuck, and if the Turing machine runs forever then a tiling of the (half) plane exists. $\square$

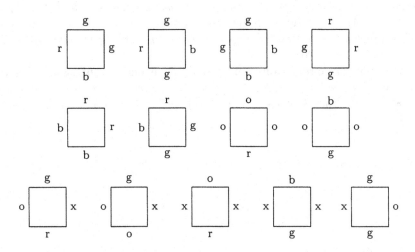

Fig. 2.18 A set of 13 tiles with 5 colors providing an aperiodic tiling of the plane as described by K. Culik II.

2.5 Combinatorial Logic

An electronic computer is composed of digital circuits that can be considered
as a network of electronically implemented logic gates. A logic gate performs
a logic operation on one or more logic inputs (e.g., 0 for low voltage and 1
for high voltage) and produces a single logic output. The logic operations
follow Boolean logic, named after G. Boole who first defined an algebraic
system of logic in the mid-19th century. Digital circuits are divided into
combinatorial and sequential circuits. Combinatorial circuits are described by
Boolean functions in which output signals depend only on the input signals,
while sequential circuits are modeled by finite state automata in which output
signals depend on the input signals and the internal states.

2.5.1 Boolean Circuits

Let $\mathbb{B} = \{0, 1\}$. An *n-ary Boolean function* is a mapping $f : \mathbb{B}^n \to \mathbb{B}$. Each
such mapping describes the input/output behavior of a combinatorial circuit
with n input signals and one output signal (Fig. 2.19). The set of all n-ary
Boolean functions $f : \mathbb{B}^n \to \mathbb{B}$ is denoted by $\mathbb{F}_n$. The number of n-ary Boolean
functions is $|\mathbb{F}_n| = 2^{2^n}$.

The 0-ary Boolean functions are the constant functions, the zero function
$0 :\mapsto 0$ and the unit function $1 :\mapsto 1$.

The unary Boolean functions are the identity mapping $id : b \mapsto b$, *negation*
$\neg : b \mapsto \neg b$, and the constant mappings $0 : b \mapsto 0$ und $1 : b \mapsto 1$ (Table 2.3).

Among the sixteen binary Boolean functions, the most important are the
conjunction $\wedge : (b_1, b_2) \mapsto b_1 \wedge b_2$ and the *disjunction* $\vee : (b_1, b_2) \mapsto b_1 \vee b_2$ (Table 2.4). The digital circuits corresponding to negation, conjunction,
and disjunction are called NOT gate or *inverter*, AND gate, and OR gate,
respectively (Fig. 2.20).

Combinatorial circuits often have more than one output signal. Let m
and n be positive integers. An *n-ary, m-adic Boolean function* is a mapping
$F : \mathbb{B}^n \to \mathbb{B}^m$. Such a Boolean function describes m combinatorial circuits,

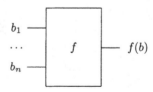

Fig. 2.19 An n-ary Boolean function f.

Table 2.3 The unary Boolean functions.

b	$id(b)$	$\neg b$	0	1
0	0	1	0	1
1	1	0	0	1

Table 2.4 Conjunction and disjunction.

b_1	b_2	$b_1 \wedge b_2$	$b_1 \vee b_2$
0	0	0	0
0	1	0	1
1	0	0	1
1	1	1	1

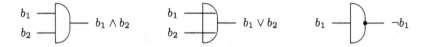

Fig. 2.20 Logic gates: AND, OR, and NOT.

each of which is given by an n-ary Boolean function operating on the same input (Fig. 2.21). Therefore, an n-ary, m-adic Boolean function is an m-tuple F of n-ary Boolean functions $f_1, \ldots, f_m$ so that

$$F(b) = (f_1(b), \ldots, f_m(b)), \quad b \in \mathbb{B}^n . \tag{2.23}$$

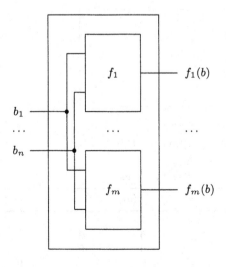

Fig. 2.21 An n-ary, m-adic Boolean function $F = (f_1, \ldots, f_m)$.

2.5.2 Compound Circuits

Let $f_1, \ldots, f_m$ be n-ary Boolean functions. Consider a digital circuit in which these Boolean functions process a common input signal in parallel and feed their output into a common m-ary Boolean function f (Fig. 2.22). This circuit is described by the n-ary Boolean function $g = f(f_1, \ldots, f_m)$ so that

$$g(b) = f(f_1(b), \ldots, f_m(b)), \quad b \in \mathbb{B}^n . \tag{2.24}$$

In particular, consider Boolean functions composed of AND, OR, and NOT gates,

$$(f_1 \wedge f_2)(b) = f_1(b) \wedge f_2(b),$$
$$(f_1 \vee f_2)(b) = f_1(b) \vee f_2(b),$$
$$(\neg f_1)(b) = \neg f_1(b), \quad b \in \mathbb{B}^n .$$

Example 2.61. Several Boolean functions composed of NAND gate $\uparrow$ and NOR gate $\downarrow$ are illustrated in Table 2.5. $\diamond$

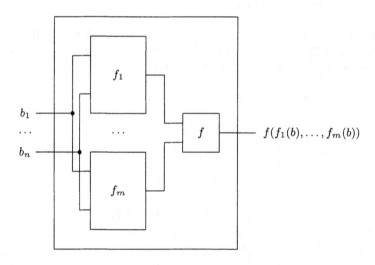

Fig. 2.22 The Boolean function $g = f(f_1, \ldots, f_m)$.

Table 2.5 Compound Boolean functions.

b_1	b_2	$\uparrow (b)$	$\downarrow (b)$	$(\neg \uparrow)(b)$	$(\uparrow \wedge \downarrow)(b)$	$(\uparrow \vee \downarrow)(b)$
0	0	1	1	0	1	1
0	1	1	0	0	0	1
1	0	1	0	0	0	1
1	1	0	0	1	0	0

Theorem 2.62. *Let n be a non-negative integer. The set of all n-ary Boolean functions forms with conjunction and disjunction a Boolean algebra $\mathbb{F}_n$. That is, for all n-ary Boolean functions f, f_1, f_2, and f_3,*

- *Commutativity:*

$$f_1 \wedge f_2 = f_2 \wedge f_1 \tag{2.25}$$
$$f_1 \vee f_2 = f_2 \vee f_1 . \tag{2.26}$$

- *Associativity:*

$$f_1 \wedge (f_2 \wedge f_3) = (f_1 \wedge f_2) \wedge f_3 \tag{2.27}$$
$$f_1 \vee (f_2 \vee f_3) = (f_1 \vee f_2) \vee f_3 . \tag{2.28}$$

- *Absorption:*

$$f_1 \wedge (f_1 \vee f_2) = f_1 \tag{2.29}$$
$$f_1 \vee (f_1 \wedge f_2) = f_1 . \tag{2.30}$$

- *Distributivity:*

$$f_1 \wedge (f_2 \vee f_3) = (f_1 \wedge f_2) \vee (f_1 \wedge f_3) \tag{2.31}$$
$$f_1 \vee (f_2 \wedge f_3) = (f_1 \vee f_2) \wedge (f_1 \vee f_3) . \tag{2.32}$$

- *Complementarity:*

$$f \wedge \neg f = 0 \tag{2.33}$$
$$f \vee \neg f = 1 . \tag{2.34}$$

In the following, an algebraic notation for the operations of the Boolean algebra $\mathbb{F}_n$ is used,

$$fg = f \wedge g, \quad f + g = f \vee g, \quad \text{and} \quad \overline{f} = \neg f . \tag{2.35}$$

An expression of Boolean functions is also termed *Boolean expression* and is evaluated according to the priorities assigned to the logical operations: Negation has the highest priority, multiplication the next highest and addition the lowest. In this way, we have $\overline{(f + g)} + (fg) = \overline{f + g} + fg$.

2.5.3 Minterms and Maxterms

The *ith projection* in $\mathbb{F}_n$ is the n-ary Boolean function

$$x_i : \mathbb{B}^n \to \mathbb{B} : b \mapsto b_i, \quad 1 \leq i \leq n . \tag{2.36}$$

The functions x_i are termed *variables*, and both x_i and $\overline{x}_i$ are called *literals*. Write

$$x_i^1 = x_i \quad \text{and} \quad x_i^0 = \overline{x}_i \tag{2.37}$$

and assign to each element $a \in \mathbb{B}^n$ two n-ary Boolean functions

$$m_a = x_1^{a_1} \cdots x_n^{a_n}, \tag{2.38}$$
$$M_a = x_1^{\overline{a}_1} + \ldots + x_n^{\overline{a}_n}. \tag{2.39}$$

Lemma 2.63. *For all elements* $a, b \in \mathbb{B}^n$,

$$m_a(b) = \begin{cases} 1 & \text{if } a = b, \\ 0 & \text{otherwise}, \end{cases} \tag{2.40}$$

and

$$M_a(b) = \begin{cases} 1 & \text{if } a \neq b, \\ 0 & \text{otherwise}. \end{cases} \tag{2.41}$$

Proof. By definition, $m_a(b) = b_1^{a_1} \ldots b_n^{a_n} = 1$ if and only if $b_i = a_i$ for all $1 \leq i \leq n$, and $M_a(b) = b_1^{\overline{a}_1} + \ldots + b_n^{\overline{a}_n} = 0$ if and only if $b_i = a_i$ for all $1 \leq i \leq n$,   □

The Boolean functions m_a and M_a, $a \in \mathbb{B}^n$, are called the *minterms* and *maxterms* of $\mathbb{F}_n$, respectively.

Example 2.64. In view of the Boolean algebra $\mathbb{F}_3$, the minterms are

$$m_0 = \overline{x}_1\overline{x}_2\overline{x}_3, \ m_3 = \overline{x}_1 x_2 x_3, \ m_6 = x_1 x_2 \overline{x}_3,$$
$$m_1 = \overline{x}_1\overline{x}_2 x_3, \ m_4 = x_1 \overline{x}_2 \overline{x}_3, \ m_7 = x_1 x_2 x_3,$$
$$m_2 = \overline{x}_1 x_2 \overline{x}_3, \ m_5 = x_1 \overline{x}_2 x_3,$$

and the maxterms are

$$M_0 = x_1 + x_2 + x_3, \ M_3 = x_1 + \overline{x}_2 + \overline{x}_3, \ M_6 = \overline{x}_1 + \overline{x}_2 + x_3,$$
$$M_1 = x_1 + x_2 + \overline{x}_3, \ M_4 = \overline{x}_1 + x_2 + x_3, \ M_7 = \overline{x}_1 + \overline{x}_2 + \overline{x}_3,$$
$$M_2 = x_1 + \overline{x}_2 + x_3, \ M_5 = \overline{x}_1 + x_2 + \overline{x}_3,$$

where the index $a = (a_2, a_1, a_0)$ corresponds to the decimal equivalent $a_2 \cdot 2^2 + a_1 \cdot 2 + a_0$.   ◇

2.5.4 Canonical Circuits

An important result in combinatorial logic is that each Boolean function can be represented by AND, OR, and NOT gates.

Theorem 2.65 (Canonical Disjunctive Normal Form, CDNF). *Each n-ary Boolean function f can be represented in the form*

$$f = \sum_{\substack{a \in \mathbb{B}^n \\ f(a)=1}} m_a \ . \tag{2.42}$$

Proof. The assertion is clear in case of $n = 0$. Let $n > 0$. Observe that each $f \in \mathbb{F}_n$ can be written in the form

$$f = x_n g + \overline{x}_n h \ ,$$

where $g, h \in \mathbb{F}_{n-1}$ so that

$$g(b_1, \ldots, b_{n-1}) = f(b_1, \ldots, b_{n-1}, 1) \ ,$$
$$h(b_1, \ldots, b_{n-1}) = f(b_1, \ldots, b_{n-1}, 0) \ .$$

By induction hypothesis,

$$g = \sum_{\substack{a \in \mathbb{B}^{n-1} \\ g(a)=1}} m_a^{(n-1)} \quad \text{and} \quad h = \sum_{\substack{a \in \mathbb{B}^{n-1} \\ h(a)=1}} m_a^{(n-1)} \ .$$

Hence,

$$\begin{aligned}
f &= x_n g + \overline{x}_n h \\
&= x_n \sum_{\substack{a \in \mathbb{B}^{n-1} \\ g(a)=1}} m_a^{(n-1)} + \overline{x}_n \sum_{\substack{a \in \mathbb{B}^{n-1} \\ h(a)=1}} m_a^{(n-1)} \\
&= \sum_{\substack{a \in \mathbb{B}^{n-1} \\ (x_n g)(a,1)=1}} x_n m_a^{(n-1)} + \sum_{\substack{a \in \mathbb{B}^{n-1} \\ (\overline{x}_n h)(a,0)=1}} \overline{x}_n m_a^{(n-1)} \\
&= \sum_{\substack{a \in \mathbb{B}^n \\ f(a)=1}} m_a \ .
\end{aligned}$$

$\square$

The following result can be similarly proved.

Theorem 2.66 (Canonical Conjunctive Normal Form, CCNF). *Each n-ary Boolean function f can be represented in the form*

$$f = \prod_{\substack{a \in \mathbb{B}^n \\ f(a)=0}} M_a \ . \tag{2.43}$$

Example 2.67. A light should be independently controlled by two switches. The corresponding combinatorial circuit has two input signals associated with the switches and one output signal for light control. Each switch can be in two positions, on (1) and off (0). Suppose at the beginning that both switches are off and the light is off. If one of the switches is pressed, the light goes on. Then if one of the switches is pressed again, the light goes off, and so on. The corresponding Boolean function f is the exclusive OR (Table 2.6), which has the CDNF $f = m_{10} + m_{01} = x_1\bar{x}_2 + \bar{x}_1 x_2$ and the CCNF $f = M_{00}M_{11} = (x_1 + x_2)(\bar{x}_1 + \bar{x}_2)$ (Fig. 2.23).   ◇

2.5.5 Adder Circuits

Adder circuits belong to the simplest Boolean functions.

Half-Adder

A *half-adder* is a combinatorial circuit for adding two bits. It can be described by a 2-ary, 2-adic Boolean function $H = (s_H, c_H)$, where s_H computes the sum and c_H provides the carry bit (Table 2.7). The half-adder has the disjunctive normal form (Fig. 2.24)

$$s_H = x_1\bar{x}_2 + \bar{x}_1 x_2 \quad \text{and} \quad c_H = x_1 x_2 . \tag{2.44}$$

Table 2.6 Exclusive OR.

b_1	b_2	$f(b)$
0	0	0
1	0	1
0	1	1
1	1	0

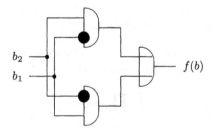

Fig. 2.23 Combinatorial circuit in DNF for light control. (The two inverters are indicated by black circles.)

Table 2.7 Half-adder.

b_1	b_2	$s_H(b)$	$c_H(b)$
0	0	0	0
0	1	1	0
1	0	1	0
1	1	0	1

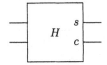

Fig. 2.24 Symbol for half-adder.

Full-Adder

A *full-adder* is a combinatorial circuit for adding three bits. It can be described by a 3-ary, 2-adic Boolean function $V = (s_V, c_V)$. The function s_V computes the sum and the function c_V provides the carry bit (Table 2.8). Clearly,

$$s_V(x_1, x_2, x_3) = s_H(x_1, s_H(x_2, x_3)), \qquad (2.45)$$
$$c_V(x_1, x_2, x_3) = c_H(x_1, s_H(x_2, x_3)) + c_H(x_2, x_3) . \qquad (2.46)$$

Therefore, a full-adder can be implemented by two half-adders and an OR gate (Fig. 2.25).

Adder

An *adder* is a combinatorial circuit for adding two n-ary binary numbers. It can be implemented by a $2n$-ary, $(n + 1)$-adic Boolean function, which is composed of n serially connected full-adders. In particular, the full-adder for the lowest bit position can be realized by a half-adder (Fig. 2.26).

Table 2.8 Full-adder.

b_1	b_2	b_3	$s_V(b)$	$c_V(b)$
0	0	0	0	0
0	0	1	1	0
0	1	0	1	0
0	1	1	0	1
1	0	0	1	0
1	0	1	0	1
1	1	0	0	1
1	1	1	1	1

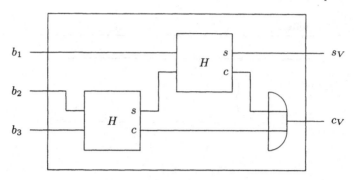

Fig. 2.25 A full-adder.

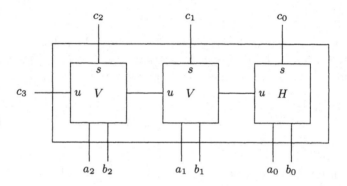

Fig. 2.26 An adder for 3-ary binary numbers.

2.6 Computational Complexity

Alan Turing's study of what could and what could not be computed was extended by S. Cook in the late 1960s. He was able to separate those problems that can be solved efficiently by computers from those problems that can in principle be solved but in practice take so much time that computers are useless for all but small problem instances.

2.6.1 Time Complexity

The *runtime* of a computer algorithm provides the duration of its execution, from beginning to termination. Runtime can be measured in milliseconds. However, this measurement depends on several parameters such as computer, compiler, operating system, programming tricks. Runtime should also be a measure for algorithms. Therefore, the runtime of a program is measured as follows:

- For each input, the number of elementary operations is counted.
- The runtime is given as a function of the number of elementary operations in dependence of the size of the input.

The *elementary operations* are assignments $(x := 3)$, comparisons $(x < y)$, arithmetic and logical operations $(x + y, x \wedge y)$, and array access $(a[i])$. Non-elementary operations are loops, conditional statements, and procedure calls.

Example 2.68. BUBBLESORT is a list-sorting algorithm, which interchanges two elements in a list if they are in the wrong order (Alg. 2.2). BUBBLESORT consists of a two-fold nested loop, which is executed for the $n(n-1)/2$ pairs

$$(2, n), (2, n-1), \ldots, (2, 2), (3, n), \ldots, (3, 3), \ldots, (n, n).$$

The loop body consists of four elementary operations. Suppose that each elementary operation has runtime t. Then the algorithm has a total runtime of $4tn(n-1)/2$. $\Diamond$

Algorithm 2.2 BUBBLESORT(L)

Input: List L of n elements
Output: List L is sorted in ascending order
1: **for** $i = 2$ to n **do**
2:    **for** $j = n$ downto i **do**
3:       **if** $L[j-1] > L[j]$ **then**
4:          $h := L[j]$
5:          $L[j] := L[j-1]$
6:          $L[j-1] := h$
7:       **end if**
8:    **end for**
9: **end for**

2.6.2 Infinite Asymptotics

The asymptotic behavior of functions is described by the Landau notation. Let $f : \mathbb{N}_0 \to \mathbb{R}_0^+$ and $g : \mathbb{N}_0 \to \mathbb{R}_0^+$ be mappings. The expression $f = O(g)$ says that the function f is *asymptotically bound* by the function g, that is,

$$\exists n_0 \in \mathbb{N} \ \exists c \in \mathbb{R}^+ \ \forall n \in \mathbb{N} \left[n \geq n_0 \Rightarrow f(n) \leq c \cdot g(n) \right]. \tag{2.47}$$

The *Landau symbol* O is called "big Oh" after E. Landau. The notation $f = O(g)$ stands for the more precise notation $f \in O(g)$, where $O(g)$ is the set of all functions f so that $f = O(g)$.

Example 2.69. • $f(n) = 3 = O(1)$, since $3 \leq 3 \cdot 1$.
• $f(n) = n + 4 = O(n)$, as $n + 4 \leq 2n$ for all $n \geq 4$.
• $f(n) = 4n + 6 = O(n)$, for $4n + 6 \leq 5n$ for all $n \geq 6$.
• $f(n) = n(n-1)/2 = O(n^2)$, because $n(n-1)/2 \leq n^2$ for all $n \geq 1$.
• $f(n) = 2n^2 + 4n + 6 = O(n^2)$, since $2n^2 + 4n + 6 \leq 3n^2$ for all $n \geq 6$.

$\diamond$

The Landau notation allows the consideration of algorithms at a coarser level at which constants are eliminated and upper bounds are formed. The big-O notation leads to a classification of algorithms via runtime (Table 2.9). To solve practical problems, algorithms with polynomial runtime $O(n^c)$, $c \geq 0$, are needed, while algorithms with exponential runtime $O(c^n)$, $c > 1$, are considered to be impractical.

Example 2.70. Suppose all elementary operations have runtime $O(1)$. Then BUBBLESORT has runtime $O(n^2)$, since the body has runtime $O(1)$ and the loop is executed $n(n-1)/2$ times. $\diamond$

Let $f : \mathbb{N}_0 \to \mathbb{R}_0^+$ and $g : \mathbb{N}_0 \to \mathbb{R}_0^+$ be mappings. The expression $f = o(g)$ says that the function g *asymptotically grows faster* than the function f. Formally, it means that the limit of $f(n)/g(n)$ is zero. That is,

$$\forall c > 0 \; \exists n_0 \in \mathbb{N} \; \forall n \in \mathbb{N} \, [n \geq n_0 \Rightarrow 0 \leq f(n) < c \cdot g(n)] . \qquad (2.48)$$

The *Landau symbol* o is called "little oh" and $o(g)$ represents the set of all functions f so that $f = o(g)$.

Example 2.71. Clearly, $\frac{1}{n} = o\left(\frac{1}{\sqrt{n}}\right)$, $n^2 = o(e^n)$, and $\log n = o(n)$. $\diamond$

Table 2.9 Common classes of functions.

Notation	Name	Typical Algorithm
$O(1)$	constant	determine if number is even or odd
$O(\log n)$	logarithmic	find item in a sorted list
$O(n)$	linear	find item in an unsorted list
$O(n \log n)$	log linear	sort a list (heap sort)
$O(n^2)$	quadratic	sort a list (bubble sort)
$O(n^c)$, $c \geq 2$	polynomial	find shortest paths in a graph
$O(2^n)$	exponential	find Hamiltonian paths in a graph

2.6.3 Decision Problems

A *decision problem* requires an answer "yes" or "no". Decision problems are divided into two classes. The class P consists of all decision problems that are solvable in polynomial time. The class NP embraces all decision problems with the property that for each problem instance answered "yes", there is a proof for the answer which can be verified in polynomial time. By definition, $P \subseteq NP$. It is conceivable that the class NP is larger than the class P. Today, this is one of the most famous open problems in Computer Science.

Example 2.72. Let $G = (V, E)$ be a graph. A subset U of V is called *independent* in G if $uv \notin E$ for all $u, v \in U$. Consider the decision problem of finding for each graph G and each integer $k > 0$ an independent set U in G so that $|U| \geq k$. Claim that this problem lies in NP. Indeed, each subset U of V has $\binom{|U|}{2}$ two-element subsets uv with $u, v \in U$. Put $n = |V|$ and so $\binom{|U|}{2} = O(n^2)$. The test, whether uv lies in E, can be performed in constant time. Thus $O(n^2)$ steps are necessary to test whether the set U is independent in G. $\diamond$

Let D and D' be decision problems. A *polynomial transformation* of D into D' is an algorithm with polynomial runtime, which transforms each instance I of D into an instance I' of D' so that the answers of both instances I and I' are identical. Write $D \propto D'$ if such a transformation exists.

Example 2.73. Let $G = (V, E)$ be a graph. A subset $U \subseteq V$ is called a *clique* in G if $uv \in E$ for all $u, v \in U$. Consider the decision problem of finding for each graph G and each integer $k > 0$ a clique U in G so that $|U| \geq k$. This problem belongs to the class NP.

Let $G' = (V, E')$ be the graph *complementary* to G (i.e., $uv \in E'$ if and only if $uv \notin E$) (Fig. 2.27). The independent sets in G are exactly the cliques in G', and vice versa. The graph G' can be derived from G in $O(|V|^2)$ steps. Hence, the decision problem of finding in a graph an independent subset with size at least k elements can be polynomially transformed into the decision problem of finding in a graph a clique with size at least k. $\diamond$

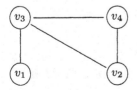 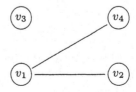

Fig. 2.27 Two complementary graphs.

A decision problem D is termed *NP-complete* if $D \in$ NP and each problem D' in NP satisfies $D' \propto D$.

Theorem 2.74. *Let D be an NP-complete decision problem. If D lies in P, then $P = NP$.*

Proof. Let $D \in$ P be an NP-complete decision problem. Let A be an algorithm solving D in polynomial time. Let $D' \in$ NP. By hypothesis, $D' \propto D$. Thus there is an algorithm B with polynomial runtime, which transforms each instance of D' into an instance of D. Therefore, the composition of the algorithms A and B solves each instance of D' in polynomial time and hence $D' \in$ P. $\square$

The satisfiability problem was the first decision problem proven to be NP-complete. Today, more than one thousand NP-complete problems are known.

Theorem 2.75 (Cook's Theorem, SAT Problem). *Let n be a non-negative integer and let f be an n-ary Boolean function. The problem of determining whether the function f is satisfiable (i.e., finding an input $(b_1, \ldots, b_n) \in \mathbb{B}^n$ so that $f(b_1, \ldots, b_n) = 1$) is NP-complete.*

Equivalently, let F be a Boolean expression in n variables. Determining whether the expression F is satisfiable (i.e., finding a truth assignment for the n variables in the expression F so that it becomes "true") is NP-complete.

Theorem 2.76 (3-SAT Problem). *Let F be a Boolean expression in n variables given in* conjunctive normal form *(CNF) (i.e., a conjunction of disjunctions (clauses) of literals). Suppose each disjunctive clause contains at most three literals such as*

$$F = (x_2 + \overline{x}_3)(x_1 + \overline{x}_2 + x_3)(\overline{x}_1 + x_2) .$$

The problem of finding a truth assignment of the variables in F so that F becomes "true" is NP-complete.

Theorem 2.77 (Vertex Cover Problem). *Let $G = (V, E)$ be a graph and let $k > 0$ be an integer. A* vertex cover *in G is a set of vertices $V' \subseteq V$ so that each edge is incident with at least one vertex in V'. The problem of finding in G a vertex cover of size at most k is NP-complete.*

Example 2.78. In view of the graph G in Figure 2.1, the sets $\{v_1, v_4\}$ and $\{v_2, v_3\}$ are vertex covers in G. $\Diamond$

Theorem 2.79 (Clique Problem). *Let G be a graph and let $k > 0$ be an integer. The problem of finding in G a clique of size at least k is NP-complete.*

Example 2.80. In view of the graph G in Figure 2.1, the set $\{v_2, v_3, v_4\}$ forms a clique in G. $\Diamond$

Theorem 2.81 (Independent Vertex Set Problem). *Let G be a graph and let $k > 0$ be an integer. The problem of finding in G an independent set of size at least k is NP-complete.*

Example 2.82. In view of the graph G in Figure 2.1, the sets $\{v_1, v_2\}$ and $\{v_1, v_4\}$ are independent in G. ◇

Theorem 2.83 (Set Cover Problem). *Let A be a set, let $\{A_1, \ldots, A_n\}$ be a set of subsets of A, and let $k > 0$ be an integer. The problem of finding a subset I of $\{1, \ldots, n\}$ so that $\bigcup_{i \in I} A_i = A$ and $|I| \leq k$ is NP-complete.*

Theorem 2.84 (Vertex Coloring Problem). *Let $G = (V, E)$ be a graph and let $k > 0$ be an integer. A* coloring *of G with k colors is a mapping $f : V \to \{1, \ldots, k\}$ which assigns colors to the vertices in G. A coloring must assign different colors to adjacent vertices. The problem of coloring G with at most $k \geq 3$ colors is NP-complete.*

Example 2.85. In view of the graph G in Figure 2.1, a 3-coloring of G is given by $f(v_1) = 1$, $f(v_2) = 1$, $f(v_3) = 2$, and $f(v_4) = 3$. ◇

Theorem 2.86 (Matching Problem). *Let $G = (V, E)$ be a graph and let $k > 0$ be an integer. A* matching *in G is a set of edges $E' \subseteq E$ so that no two edges in E' have a vertex in common. The problem of finding in G a matching of size at least k is NP-complete.*

Example 2.87. In view of the graph G in Figure 2.1, a matching in G is given by $\{v_1 v_3, v_2 v_4\}$. This matching is perfect. ◇

Theorem 2.88 (Perfect Matching Problem). *Let $n > 0$ be an even integer, and let G be a graph with n vertices. A* perfect matching *in G is a matching in G with $n/2$ edges. The problem of finding in G a perfect matching is NP-complete.*

Theorem 2.89 (Edge-Dominating Set Problem). *Let $G = (V, E)$ be a graph and let $k > 0$ be an integer. For each subset of edges $E' \subseteq E$, let $n_G(E')$ be the set of all edges in G which have at least one vertex in common with an edge in E'. If $n_G(E') = E$, then E' is called an* edge-dominating set *in G. The problem of finding in G an edge-dominating set of size at most k is NP-complete.*

Example 2.90. In view of the graph G in Figure 2.1, edge-dominating sets in G are $\{v_2 v_4\}$ and $\{v_3 v_4\}$. ◇

Theorem 2.91 (Bipartite Subgraph Problem). *Let G be a graph and let $k > 0$ be an integer. The problem of finding in G a bipartite subgraph with at least k edges is NP-complete.*

Theorem 2.92 (Hamiltonian Path Problem). *Let G be a graph. A simple path in G is called* Hamiltonian *if it contains each vertex of G. The problem of finding in G a Hamiltonian path is NP-complete.*

Example 2.93. The graph G in Figure 2.1 exhibits two Hamiltonian paths, (v_1, v_3, v_2, v_4) and (v_1, v_3, v_4, v_2).   ◇

Theorem 2.94 (Hamiltonian Cycle Problem). *Let G be a graph. A simple cycle in G is called* Hamiltonian *if it contains each vertex of G. The problem of finding in G a Hamiltonian cycle is NP-complete.*

Theorem 2.95 (Steiner Tree Problem). *Let $k > 0$ be an integer, let $G = (V, E)$ be a connected graph and let Z be a subset of V. A subtree $H = (U, F)$ of G is called a* Steiner tree *for Z in G if $Z \subseteq U$ and H has a minimal number of edges among all subtrees of G that contain the vertices of Z. In the case of $Z = V$, a Steiner tree in G is a spanning tree in G.*

The problem of finding a Steiner tree $H = (U, F)$ for Z in G so that $k \geq |F|$ is NP-complete. The problem of finding a Steiner tree $H = (U, F)$ for Z in G so that $k \geq |U \setminus Z|$ is NP-complete.

2.6.4 Optimization Problems

Combinatorial problems often arise in the form of optimization problems. These problems can be classified by using a reduction technique. A *Turing reduction* assigns to each decision problem D an optimization problem D' so that a decision algorithm A for D is used as a "subalgorithm" in a solution algorithm B for D'. Additionally, B should have polynomial runtime if A has polynomial runtime. Such a reduction is denoted by $D' \propto_T D$.

Example 2.96. Consider the decision problem of finding a clique of size at least k in a graph of order n. The corresponding optimization problem aims at finding a clique of maximum size in a graph. Let A be an algorithm solving the decision problem. This algorithm can be used in a loop, which computes in each step from $k = n$ down to $k = 1$ an instance $A(G, k)$ of the decision problem. As soon as an instance $A(G, k)$ provides the answer "yes", a maximal clique is found and the optimization problem is solved.   ◇

The analogue of NP-complete decision problems are NP-hard optimization problems. A problem D is called *NP-hard* if there is an NP-complete problem D' so that $D' \propto_T D$. In Example 2.96, the optimization problem of finding a maximal clique in a graph is NP-hard. In general, the optimization problems corresponding to NP-complete decision problems are NP-hard.

References

1. Cook S (1971) The complexity of theorem-proving procedures. Proc 3rd Ann ACM Symp Theor Comp 151–158
2. Culik II K (1996) An aperidic set of 13 Wang tiles. Disc Math 160:245–251

3. Garey MR, Johnson DS (1979) Computers and intractability: a guide to the theory of NP-completeness. Freeman, New York
4. Hopcraft E, Motwani R, Ullman JD (2007) Automata theory, languages, and computation. Addison Wesley, Boston
5. Mendelson E (1970) Schaum's outline of theory and problems of Boolean algebra and switching circuits. McGraw-Hill, London
6. Schönig U (1997) Theoretische Informatik – kurzgefasst. Spektrum, Heidelberg Berlin
7. Turing AM (1937) On computable numbers, with an application to the Entscheidungsproblem. Proc London Math Soc Ser 42:230–265
8. Wang H (1961) Proving theorems by pattern recognition II. Bell Syst Tech J 40:1–41
9. Wolfram S (1983) Theory and application of cellular automata. Addison Wesley, Reading, MA
10. Wolfram S (2002) A new kind of science. Wolfram Media, Champaign, IL

Chapter 3
Molecular Biology

Abstract Genetic information is passed with high accuracy from the parental organism to the offspring and its expression governs the biochemical and physiological tasks of the cell. Although different types of cells exist and are shaped by development to fill different physiological niches, all cells have fundamental similarities and share common principles of organization and biochemical activities. This chapter gives an overview of general principles of the storage and flow of genetic information. It aims to summarize and describe in a broadly approachable way, from the point of view of molecular biology, some general terms, mechanisms and processes used as a base for the molecular computing in the subsequent chapters.

3.1 DNA

In the majority of living organisms the genetic information is stored in the desoxyribonucleic acid (DNA), molecules that govern the development and functions of the organisms. The high accuracy of duplication and transmission of the DNA is determined by its structural features and the unique fidelity of the proteins participating in this process.

3.1.1 Molecular Structure

DNA is composed of four nucleotides, also called *bases:* adenosine (**A**), cytidine (**C**), guanosine (**G**), and thymidine (**T**), each of which consists of a phosphate group, a sugar (deoxyribose), and a nucleobase (pyrimidine – thymine and cytosine, or purine – adenine and guanine). The nucleotides are covalently linked through the sugar (deoxyribose) and phosphate residue and form the backbone of one DNA strand (Fig. 3.1). These two different elements

Z. Ignatova et al., *DNA Computing Models,*
DOI: 10.1007/978-0-387-73637-2_3, © Springer Science+Business Media, LLC 2008

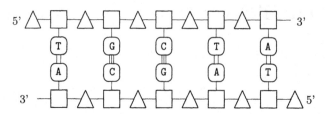

Fig. 3.1 Schematic overview of the DNA structure. The phosphate group is shown as a triangle, the sugar component is depicted as a square and together they form the backbone. The double helix is stabilized by hydrogen bonds between A and T (two hydrogen bonds) and G and C (three hydrogen bonds).

(sugar and the phosphate group) alternate in the backbone and determine the *directionality* of the DNA: the end with the exposed hydroxyl group of the deoxyribose is known as the 3' end; the other end with the phosphate group is termed the 5' end.

Two single DNA strands assemble into a double-stranded DNA molecule, which is stabilized by hydrogen bonds between the nucleotides. The chemical structure of the bases allows an efficient formation of hydrogen bonds only between A and T or G and C; this determines the *complementarity principle*, also known as *Watson-Crick base-pairing* of the DNA double helix. The A and T base pair aligns through a double hydrogen bond and the G and C pair glues with a triple hydrogen bond, which is the reason for the higher stability of the G–C *Watson-Crick base pair* over the A–T *Watson-Crick base pair*. The overall stability of the DNA molecule increases with the increase of the proportion of the G–C base pairs. The two single DNA strands are complementarily aligned in a reverse direction: the one, called also a *leading strand*, has a 5' to 3' orientation, whereas the complementary strand, called *lagging strand*, is in the reverse 3' to 5' orientation (Fig. 3.1).

In aqueous solution the two single strands wind in an anti-parallel manner around the common axis and form a twisted right-handed double helix with a diameter of about 20 Å. The planes of the bases are nearly perpendicular to the helix axis and each turn accommodates 10 bases. The wrapping of the two strands around each other leads to a formation of two grooves: the major is 22 Å wide and the minor is 12 Å wide. This structure is known as B-DNA and represents the general form of the DNA within the living cells. Alternatively the DNA double helix can adopt several other conformations (e.g., A-DNA and Z-DNA), which differ from the B-form in their dimensions and geometry. Unlike the A-DNA which is also right-handed, the Z-DNA is left-handed and the major and minor grooves show differences in width. The propensity to adopt one of these alternative conformations depends on the sequence of the polynucleotide chain and the solution conditions (e.g., concentrations of metal ions, polyamines). The hybrid pairing of DNA and RNA strands has under physiological conditions an A-form like conformation, while the Z-form is believed to occur during transcription of the DNA, providing a torsional relief for the DNA double helix.

The concept of the DNA complementarity is crucial for its functionality and activity. The base-pairing can be reversibly broken, which is essential for the DNA replication. The non-covalent forces that stabilize the DNA double helix can be completely disrupted by heating. The collapse of the native structure and the dissociation of the double helix into two single strands is called *denaturation* (Fig. 3.2). Under slow decrease in temperature, the correct base pairing can be established again and the DNA renatures. The process of binding of two single strands and the formation of a double strand is known as *annealing* or *hybridization*. The annealing conditions need to be established by a slow change of temperature, as a rapid decrease in temperature forces a fast renaturation and results in both intramolecular (within one strand) and intermolecular (within different strands) base-pairing (Fig. 3.2). Complementary stretches within one single strand that are in close proximity can re-associate to partial double-stranded intramolecular structures, known as *foldback structures*.

The DNA double helix is very stable; the entire network of hydrogen bonds and hydrophobic interactions between the bases is responsible for its global stability. Nevertheless, each single hydrogen bond is weak and short stretches from the double-stranded DNA can even be opened at physiological temperature with the help of *initiation proteins*. Each strand in the DNA serves as a template for the replication machinery, with the DNA polymerase

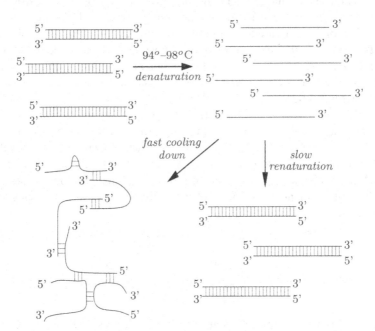

Fig. 3.2 Denaturation and renaturation of DNA.

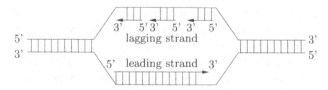

Fig. 3.3 Replication of DNA.

as a major player, replicating them in a complementary manner (Fig. 3.3). The DNA replication is asymmetric and DNA polymerase elongates in the 5' to 3' direction only. The opposite DNA strand is discontinuously synthesized again in the 5' to 3' direction as small fragments, called *Okazaki fragments*. The Okazaki fragments are further covalently joined by DNA ligase. In each replication cycle the double-stranded DNA template is replicated into two identical copies.

3.1.2 Manipulation of DNA

In DNA computing, DNA is utilized as a substrate for storing information. Depending on the model of DNA computation, information is stored in the form of single-stranded DNA and/or double-stranded DNA molecules. This stored information can be manipulated by enzymes. One class of enzymes, *restriction endonucleases*, recognizes a specific short sequence of DNA, called *restriction site*, and cuts the covalent bonds between the adjacent nucleotides (Fig. 3.4). Restriction fragments are generated with either cohesive or sticky ends or blunt ends.

DNA ligase covalently links the 3' hydroxil end of one nucleotide with 5' phosphate end of another, thus repairing backbone breaks (Fig. 3.5). The *exonucleases* are enzymes that hydrolyze phosphodiester bonds from either the 3' or 5' terminus of single-stranded DNA or double-stranded DNA molecules and remove residues one at a time. Endonucleases can cut individual covalent bonds within the DNA molecules, generating discrete fragments.

BamHI	SmaI
5′ − G|GATCC − 3′	5′ − CCC|GGG − 3′
3′ − CCTAG|G − 5′	3′ − GGG|CCC − 5′

Fig. 3.4 Restriction sites of BamHI and SmaI. The restriction enzymes recognize palindrome sequences with a two-fold rotational symmetry.

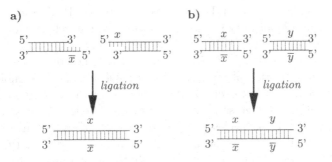

Fig. 3.5 Ligase connects **a** sticky or **b** blunt ends of double-stranded DNA.

Polymerase Chain Reaction

The template DNA can be amplified in a *polymerase chain reaction* (PCR)
(Fig. 3.6). PCR is based on the interaction of DNA polymerase with DNA.
PCR is an iterative process, with each iteration consisting of the following
steps: annealing of the short single-stranded DNA molecules, called *primers*,
that complementary pair of the templates' ends; extending of the primers in
the 5' to 3' direction by DNA polymerase by successively adding nucleotides
to the 3' end of the primer; denaturating of the newly elongated double-
stranded DNA molecules to separate their strands; and cooling to allow
re-annealing of the short, newly amplified single-stranded DNAs. Each cycle

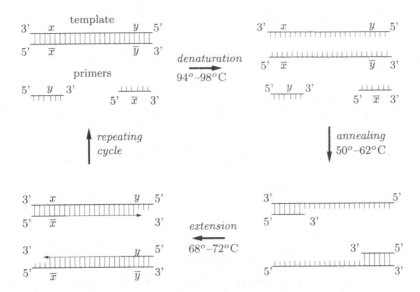

Fig. 3.6 One cycle of PCR.

doubles the number of target DNA molecules. Today, PCR is one of the most fundamental laboratory techniques in modern molecular biology. PCR is based on the interaction of DNA polymerase with DNA.

Parallel overlap assembly (POA) is a method to generate a pool of DNA molecules (combinatorial library). Short single-stranded DNA molecules, called also *oligonucleotides*, overlap after annealing and their sticky ends are extended by DNA polymerase in 5' to 3' direction. Repeated denaturation, annealing, and extension cycles increase the length of the strands. Unlike PCR, where the target DNA strands double in every cycle, in POA, the number of DNA strands does not change, only the length increases with the cycle progression (Fig. 3.7).

The short, single-stranded DNA molecules (oligonucleotides) can be designed by using available software (e.g., DNASequenceGenerator). The GC-contents can be specified as input affecting the melting temperature of the sequences. Oligonucleotides can be synthesized in vitro using PCR.

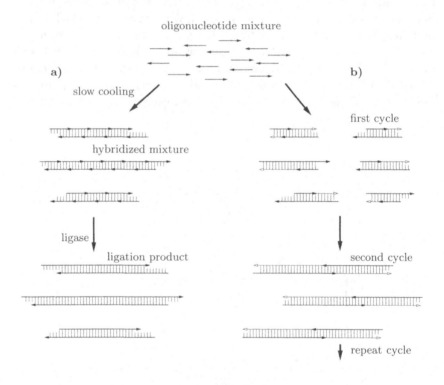

Fig. 3.7 Synthesis methods for combinatorial libraries: **a** Annealing/ligation: The arrow heads indicate the 3' end. **b** POA: The thick arrow represents the synthesized oligomers which are the input of the computation. The thin arrows represent the part that is elongated by DNA polymerase. The arrow heads indicate the 3' ends.

3.2 Physical Chemistry

Computing with biological macromolecules such as DNA is based on a fundamental physicochemical process. Therefore, knowledge about the thermodynamics and kinetics of these processes is necessary.

3.2.1 Thermodynamics

The thermodynamics of physicochemical processes is concerned with energy changes accompanying physical and chemical changes. This section addresses the thermodynamics of DNA pairing and denaturation of DNA molecules.

Nearest Neighbor Model

The relative stability of a double-stranded DNA molecule appears to depend primarily on the identity of the nearest neighbor bases. Ten different nearest neighbor interactions are possible in any double-stranded DNA molecule. These pairwise interactions are AA/TT, AT/TA, TA/AT, CA/GT, GT/CA, CT/GA, GA/CT, CG/GC, GC/CG, and GG/CC, denoted in the direction of 5' to 3'/3' to 5'. The relative stability and temperature-dependent behavior of each DNA nearest neighbor interaction can be characterized by Gibbs free energy, enthalpy, and entropy. *Gibbs free energy* describes the potential of a reaction to occur spontaneously; *enthalpy* provides the amount of heat released from or absorbed by the system; and *entropy* measures the randomness or disorder of a system. The corresponding parameters presented in Table 3.1 were derived from J. SantaLucia, Jr. and D. Hicks (2004) in 1 M NaCl at temperature 37°C. The Gibbs free energy $\Delta G°$ is related to the enthalpy $\Delta H°$ and the entropy $\Delta S°$ by the standard thermodynamic relationship

$$\Delta G° = \Delta H° - T\Delta S° .$$

(3.1)

As the Gibbs free energy data listed in Table 3.1 were calculated at 37°C, the $\Delta G°$ values at any other temperature can be computed by using the tabulated enthalpy and entropy data. For instance, the relative stability of the GC/CG pair at 50°C is $(-9.8\,\mathrm{kcal}) - [(323.15\,\mathrm{K})(-0.0244\,\mathrm{kcal/K})] = -1.915\,\mathrm{kcal}$ per mol compared with $-2.24\,\mathrm{kcal/mol}$ at 37°C.

Gibbs Free Energy

The Gibbs free energy of a double-stranded molecule given by $x = a_1 \ldots a_n$, with reverse complementary strand $\bar{a}_n \ldots \bar{a}_1$, is calculated as

Table 3.1 Nearest neighbor thermodynamics. The units for $\Delta G°$ and $\Delta H°$ are kcal/mol of interaction, and the unit for $\Delta S°$ is cal/K per mol of interaction. The symmetry correction is applied only to self-complementary duplexes. The terminal AT penalty applies to each end of a duplex that has terminal AT. A duplex with both ends closed by AT pairs has a penalty of $+1.0$ kcal/mol for $\Delta G°$.

Interaction	$\Delta H°$	$\Delta S°$	$\Delta G°$
AA/TT	−7.6	−21.3	−1.00
AT/TA	−7.2	−20.4	−0.88
TA/AT	−7.2	−21.3	−0.58
CA/GT	−8.5	−22.7	−1.45
GT/CA	−8.4	−22.4	−1.44
CT/GA	−7.8	−21.0	−1.28
GA/CT	−8.2	−22.2	−1.30
CG/GC	−10.6	−27.2	−2.17
GC/CG	−9.8	−24.4	−2.24
GG/CC	−8.0	−19.9	−1.84
Initiation	+0.2	−5.7	+1.96
Terminal AT penalty	+2.2	+6.9	+0.05
Symmetry correction	0.0	−1.4	+0.43

$$\Delta G°(x) = \Delta g_i + \Delta g_s + \sum_{i=1}^{n-1} \Delta G°(a_i a_{i+1}/\bar{a}_i \bar{a}_{i+1}) , \qquad (3.2)$$

where Δg_i denotes the helix-initiation energy and Δg_s is the symmetry correction.

Example 3.1. Consider the double-stranded DNA molecule

$$5' - \text{GCAATGGC} - 3'$$
$$3' - \text{CGTTACCG} - 5' .$$

The Gibbs free energy is given by

$$\begin{aligned}
\Delta G° &= \Delta g_i + \Delta g_s + \Delta G°(\text{GC/CG}) + \Delta G°(\text{CA/GT}) + \Delta G°(\text{AA/TT}) \\
&\quad + \Delta G°(\text{AT/TA}) + \Delta G°(\text{TG/AC}) + \Delta G°(\text{GG/CC}) + \Delta G°(\text{GC/CG}) \\
&= 1.96 + 0.0 - 2.24 - 1.45 - 1.00 - 0.88 - 1.44 - 1.84 - 2.24 \\
&= -9.13 \text{ kcal/mol.}
\end{aligned}$$

The enthalpy of a double-stranded molecule given by $x = a_1 \ldots a_n$, with reverse complementary strand $\bar{a}_n \ldots \bar{a}_1$, is computed as

$$\Delta H°(x) = \Delta h_i + \sum_{i=1}^{n-1} \Delta H°(a_i a_{i+1}/\bar{a}_i \bar{a}_{i+1}) , \qquad (3.3)$$

where Δh_i denotes the helix initiation enthalpy. The entropy of a short stretch of double-stranded DNA, also called *duplex*, can either be computed from Table 3.1 or by using Eq. (3.1).

Example 3.2. The double-stranded DNA molecule in the above example has the enthalpy $\Delta H° = -59.2\,\text{kcal/mol}$ and the entropy $\Delta S° = -161.5\,\text{cal/K}$ per mol. Alternatively, the entropy at $T = 37°C$ can be calculated as

$$\Delta S° = \frac{(\Delta H° - \Delta G°) \cdot 1000}{T} = \frac{(-59.2 + 9.13) \cdot 1000}{310.15} = 161.4\,\text{cal/K per mol.}$$

Melting Temperature

The *melting temperature* is the temperature at which half of the strands in a solution are complementary base-paired and half are not. Melting is the opposite process of hybridization, which is the separation of double strands into single strands. When the reaction temperature increases, an increasing percentage of double strands melt. For oligonucleotides in solution, the melting temperature is given by

$$T_m = \frac{\Delta H°}{\Delta S° + R\ln([C_T]/z)}, \qquad (3.4)$$

where R is the gas constant, $[C_T]$ is the total molar strand concentration, and z equals 4 for nonself-complementary strands and equals 1 for self-complementary strands. Melting curves can be measured by UV absorbance at 260 nm. With the temperature, the amount of dsDNA decreases (which is paralleled by increase of the amount of ssDNA) and leads to enhancement of the absorbance at 260 nm.

Example 3.3. In view of the above non-self-complementary duplex with strand concentration of 0.2 mM for each strand, the melting temperature is

$$T_m = \frac{-59.2 \cdot 1000}{-161.5 + 1.987 \cdot \ln(0.0004/4)} - 273.15°C = 56.1°C.$$

3.2.2 Chemical Kinetics

The kinetics of physicochemical processes is concerned with the reaction rates of the reactants. This section addresses specific reactions involving DNA molecules.

Chemical Reactions

A chemical reaction is a process that results in an interconversion of chemical substances. The substances initially involved in a chemical reaction are termed *reactants*. Chemical reactions are usually characterized by a chemical change, and they provide one or more products which are generally different from the reactants.

Consider a spatially homogeneous mixture of m reactants X_i, $1 \leq i \leq n$, which react to provide a mixture of n products Y_j, $1 \leq j \leq l$. This chemical reaction can be formally described by the chemical equation

$$\alpha_1 X_1 + \ldots + \alpha_n X_n \xrightarrow{k} \beta_1 Y_1 + \ldots + \beta_l Y_l , \tag{3.5}$$

where α_i and β_j are the stoichiometric coefficients with respect to X_i and Y_j. This reaction states that α_1 molecules of substance X_1 react with α_2 molecules of substance X_2 and so on, to give β_j molecules of substance Y_j, $1 \leq j \leq l$. The reaction (3.5) can be described by the *reaction-rate equation*

$$r = k[X_1]^{\alpha_1} \cdots [X_n]^{\alpha_n}, \tag{3.6}$$

where r is the *reaction rate* (in M/s), k is the *rate constant*, and $[X_i]$ is the concentration (in mol/l) of the reactant X_i. The rate constant $k = k(T)$ is mainly affected by the reaction temperature T as described by the *Arrhenius equation*

$$k = \kappa e^{-E_a/RT} , \tag{3.7}$$

where κ is the frequency collision factor, E_a is the activation energy (in kcal/mol) necessary to overcome so that the chemical reaction can take place, and R is the gas constant.

The *order* of a chemical reaction is the power to which its concentration term is raised in the reaction-rate equation. Hence, the order of the reaction (3.5) is given by the term $\alpha = \sum_i \alpha_i$. Generally, reaction orders are determined by experiments. For instance, if the concentration of reactant X_i is doubled and the rate increases by 2^{α_i}, then the order of this reactant is α_i. In view of (3.6), the unit of the reaction constant is $(M/s)/M^\alpha$, where α is the reaction order.

Deterministic Chemical Kinetics

The traditional way of treating chemical reactions in a mathematical manner is to translate them into ordinary differential equations. Suppose that there are sufficient molecules so that the number of molecules can be approximated as a continuously varying quantity that varies deterministically over time. Then a chemical reaction can be described by a coupled system of differential equations for the concentrations of each substance in terms of the concentrations of all others:

$$\frac{d[X_i]}{dt} = f_i([X_1], \ldots, [X_n]), \quad 1 \le i \le n \,. \tag{3.8}$$

Subject to prescribed initial conditions, these differential reaction-rate equations can only be solved analytically for rather simple chemical systems. Alternatively, these systems can be tackled numerically by using a finite difference method.

Example 3.4. The *Lokta-Volterra system* describes a set of coupled autocatalytic reactions:

$$\overline{X} + Y_1 \xrightarrow{c_1} 2Y_1,$$
$$Y_1 + Y_2 \xrightarrow{c_2} 2Y_2,$$
$$Y_2 \xrightarrow{c_3} Z \,.$$

Here the bar over the reactant X signifies that its molecular population level is assumed to remain constant. These reactions also mathematically model a simple predator-prey ecosystem. The first reaction describes how prey species Y_1 reproduces by feeding on foodstuff X; the second reaction explains how predator species Y_2 reproduces by feeding on prey species Y_1; and the last reaction details the eventual demise of predator species Y_2 through natural causes. The corresponding reaction-rate equations are as follows:

$$\frac{d[Y_1]}{dt} = c_1[\overline{X}][Y_1] - c_2[Y_1][Y_2],$$
$$\frac{d[Y_2]}{dt} = c_2[Y_1][Y_2] - c_3[Y_2] \,.$$

$\Diamond$

Example 3.5. Consider the first-order reaction (e.g., irreversible isomerization or radioactive decay),

$$X \xrightarrow{k} Y \,.$$

The corresponding differential reaction-rate equation is given by

$$\frac{d[X]}{dt} = -k[X] \,.$$

In view of the initial condition $[X] = X_0$ at $t = 0$, the solution is

$$[X](t) = X_0 e^{-kt} \,.$$

3.2.3 DNA Annealing Kinetics

DNA annealing kinetics describes the reversible chemical reaction of the annealing of complementary single-stranded DNA into double-stranded DNA.

DNA pairing from single-stranded oligonucleotides is described by the chemical equation

$$\text{ssDNA}_1 + \text{ssDNA}_2 \underset{k_r}{\overset{k_f}{\rightleftharpoons}} \text{dsDNA} . \tag{3.9}$$

This reaction can proceed in both directions and thus is reversible. The forward (k_f) and reverse (k_r) rate constants describe the forward hybridization reaction and the reverse denaturation reaction, respectively. When the reaction (3.9) reaches the equilibrium, both forward and reverse reaction rates are equal. Then the concentrations are constant and do not change with time.

The forward rate constant k_f depends on DNA length, sequence context, and salt concentration:

$$k_f = \frac{k'_N \sqrt{L_s}}{N} , \tag{3.10}$$

where L_s is the length of the shortest strand participating in the duplex formation; N is the total number of base pairs present; and k'_N is the nucleation rate constant, estimated to be $(4.35 \log_{10}[\text{Na}^+] + 3.5) \times 10^5$ where $0.2 \leq [\text{Na}^+] \leq 4.0 \, \text{mol/l}$. The reverse rate constant k_r is very sensitive to DNA length and sequence:

$$k_r = k_f e^{\Delta G^\circ / RT} , \tag{3.11}$$

where R is the gas constant and T is the incubation temperature. Hybridization in vitro is usually carried out at temperature $T = T_m - 298.15 \, \text{K}$, where T_m denotes the melting temperature.

3.2.4 Strand Displacement Kinetics

DNA kinetics has the specific feature that displacement of DNA strands can take place. This is described by the chemical equation

$$A_m/B + B_m \underset{k_r}{\overset{k_f}{\rightleftharpoons}} A_m/B_m + B , \tag{3.12}$$

where A_m/B stands for a partially double-stranded DNA molecule; B_m stands for an oligonucleotide; A_m/B_m is the resulting completely complementary (perfectly paired) double-stranded DNA molecule; and B is the released single DNA strand (Fig. 7.26). This second-order reaction can be described by the differential reaction-rate equation

$$\frac{d[A_m/B_m]}{dt} = \frac{d[B]}{dt} = k_f[A_m/B][B_m] - k_r[A_m/B_m][B] . \tag{3.13}$$

The concentration of A_m/B at time t depends on its initial concentration $[A_m/B]_0$ and the concentration of A_m/B_m at time t,

$$[A_m/B] = [A_m/B]_0 - [A_m/B_m] . \tag{3.14}$$

Similarly,

$$[B_m] = [B_m]_0 - [A_m/B_m] . \tag{3.15}$$

Under appropriate hybridization conditions, the dissociation rate constant k_r of the reverse reaction is neglible. In this way, we obtain

$$\frac{d[A_m/B_m]}{dt} = k_f([A_m/B]_0 - [A_m/B_m])([B_m]_0 - [A_m/B_m]) . \tag{3.16}$$

Equivalently, we have

$$\int \frac{d[A_m/B_m]}{([A_m/B]_0 - [A_m/B_m])([B_m]_0 - [A_m/B_m])} = k_f \int dt . \tag{3.17}$$

Integration yields the concentration of the product A_m/B_m at time t,

$$[A_m/B_m] = \frac{[A_m/B]_0[B_m]_0(1 - e^{([B_m]_0 - [A_m/B]_0)k_f t})}{[A_m/B]_0 - [B_m]_0 e^{([B_m]_0 - [A_m/B]_0)k_f t}} . \tag{3.18}$$

This equation shows that at large time instant t, that is, after the reaction is complete, the concentration of the product A_m/B_m tends towards the concentration of the reactant, either A_m/B or B_m, depending on which of the initial concentrations is lower.

3.2.5 Stochastic Chemical Kinetics

Deterministic chemical kinetics assumes that a chemical reaction system evolves continuously and deterministically over time. But this process is neither continuous, as the molecular population level can change only by a discrete integer amount, nor deterministic, as it is impossible to predict the exact molecular population levels at future time instants without taking into account positions and velocities of the molecules.

Master Equations

In view of the shortcomings of deterministic chemical kinetics, the time evolution of a chemical system can be alternatively analyzed by a kind of random-walk process. This process can be described by a single differential-difference equation known as a master equation.

Suppose that there is a container of volume V containing a spatially uniform mixture of n chemical substances which can interact through m specific chemical reactions. This chemical system can be represented by the probability density function $P(X_1, \ldots, X_n; t)$, which denotes the probability that there will be X_i molecules of the ith substance in volume V at time t, $1 \leq i \leq n$. The knowledge of this function would provide a complete stochastic characterization of the system at time t. In particular, the kth *moment* of the probability density function P with respect to X_i, $1 \leq i \leq n$, is given by

$$X_i^{(k)}(t) = \sum_{X_1=0}^{\infty} \cdots \sum_{X_n=0}^{\infty} X_i^k P(X_1, \ldots, X_n; t), \quad k \geq 0. \qquad (3.19)$$

The first and second moments are of special interest. While the *mean* $X_i^{(1)}(t)$ provides the average number of molecules of the ith substance in volume V at time t, the *root-mean-square deviation* that occurs about this average is given by

$$\Delta_i(t) = \sqrt{X_i^{(2)}(t) - [X_i^{(1)}(t)]^2}. \qquad (3.20)$$

In other words, we may expect to find between $X_i^{(1)}(t) - \Delta_i(t)$ and $X_i^{(1)}(t) + \Delta_i(t)$ molecules of the ith substance in volume V at time t.

The master equation describes the time evolution of the probability density function $P(X_1, \ldots, X_n; t)$. For this, let $a_\mu dt$ denote the probability that an R_μ reaction will occur in volume V during the next time interval of length dt given that the system is in state $(X_1, \ldots, X_n)$ at time t, $1 \leq \mu \leq m$. Moreover, let $b_\mu dt$ denote the probability that the system undergoes an R_μ reaction in volume V during the next time interval of length dt, $1 \leq \mu \leq m$. Then the time evolution of the chemical system can be described by the *master equation*

$$P(X_1, \ldots, X_n; t + dt) = P(X_1, \ldots, X_n; t)[1 - \sum_{\mu=1}^{m} a_\mu dt] + \sum_{\mu=1}^{m} b_\mu dt.$$

$$(3.21)$$

The first term is the probability that the system will be in the state $(X_1, \ldots, X_n)$ at time t and will remain in this state during the next time interval of length dt. The second term provides the probability that the system undergoes at least one R_μ reaction during the next time interval of length dt, $1 \leq \mu \leq m$. The master equation can be equivalently written as

$$\frac{\delta}{\delta t} P(X_1, \ldots, X_n; t) = \sum_{\mu=1}^{m} [b_\mu - a_\mu P(X_1, \ldots, X_n; t)] . \qquad (3.22)$$

The probability density $a_\mu dt$ can be expressed by another probability density. For this, let h_μ be the random variable which specifies the number of distinct molecular reactant combinations for the reaction R_μ present in volume V at time t, $1 \leq \mu \leq m$ (Table 3.2). Moreover, let c_μ be the so-called *stochastic reaction constant* depending only on the physical properties of the molecules and the temperature of the system, so that $c_\mu dt$ is the average probability that a particular combination of R_μ reactant molecules will react in the next time interval of length dt, $1 \leq \mu \leq m$. Thus,

$$a_\mu dt = h_\mu c_\mu dt, \quad 1 \leq \mu \leq m . \qquad (3.23)$$

The stochastic reaction constants depend on the type of chemical reaction. To this end, notice that a chemical reactant X has $x = N_A[X]V$ molecules in a volume of V litres, where N_A is the Avogadro number. For instance, the first-order reaction $X \xrightarrow{k} Y$ amounts to the reaction-rate equation $r = k[X]$ M/s. The reaction decreases X at a rate of $N_A k[X]V = kx$ molecules per second and delivers cx molecules per second. Thus, $c = k$. The second-order reaction $X + Y \xrightarrow{k} Z$ gives rise to the reaction-rate equation $r = k[X][Y]$ M/s. The reaction proceeds at a rate of $N_A k[X][Y]V = kxy/(N_A V)$ molecules per second and provides cxy molecules per second. Thus, $c = k/(N_A V)$.

Example 3.6. Reconsider the Lokta-Volterra system in Example 3.4. The corresponding master equation is given by

$$\frac{\delta}{\delta t} P = c_1[(Y_1 - 1)P(X, Y_1 - 1, Y_2, Z; t) - Y_1 P]$$
$$+ c_2[(Y_1 + 1)(Y_2 - 1)P(X, Y_1 + 1, Y_2 - 1, Z; t) - Y_1 Y_2 P]$$
$$+ c_3[(Y_2 + 1)P(X, Y_1, Y_2 + 1, Z - 1; t) - Y_2 P] ,$$

Table 3.2 Reaction R_μ and corresponding random variable h_μ.

Reaction R_μ	Random Variable h_μ
$\emptyset \rightarrow$ products	$h_\mu = 1$
$A_1 \rightarrow$ products	$h_\mu = X_1$
$A_1 + A_2 \rightarrow$ products	$h_\mu = X_1 X_2$
$2A_1 \rightarrow$ products	$h_\mu = X_1(X_1 - 1)/2$
$A_1 + A_2 + A_3 \rightarrow$ products	$h_\mu = X_1 X_2 X_3$
$A_1 + 2A_2 \rightarrow$ products	$h_\mu = X_1 X_2(X_2 - 1)/2$
$3A_1 \rightarrow$ products	$h_\mu = X_1(X_1 - 1)(X_1 - 2)/6$

where $P = P(X, Y_1, Y_2, Z; t)$ is assumed to be independent of reactant X. $\diamond$

Example 3.7. Consider the first-order reaction

$$X \xrightarrow{c} Y .$$

The corresponding master equation has the form

$$\frac{\delta}{\delta t} P(X; t) = c[(1 - \delta_{X_0, X})(X + 1)P(X + 1; t) - XP(X; t)] .$$

In view of the initial condition $P(X; 0) = \delta_{X, X_0}$, the master equation can be solved analytically yielding the standard binomial probability function

$$P(X; t) = \binom{X_0}{X} e^{-cXt}[1 - e^{-ct}]^{(X_0 - X)}, \quad 0 \leq X \leq X_0 .$$

The mean of the probability function is given by

$$X^{(1)}(t) = X_0 e^{-ct}$$

and the root-mean-square deviation turns out to be

$$\Delta(t) = \sqrt{X_0 e^{-ct}(1 - e^{-ct})} .$$

$\diamond$

The master equation is mathematically tractable only for simple chemical systems. Fortunately, there is a way to evaluate the time behavior of a chemical system without having to deal with the master equation directly.

Gillespie's Direct Reaction Method

Suppose the chemical system is in state $(X_1, \ldots, X_n)$ at time t. In order to drive the system forward, two questions must be answered: When will the next reaction occur? What kind of reaction will occur?

To this end, consider the so-called *reaction probability density function* $P(\tau, \mu)$ so that $P(\tau, \mu)d\tau$ is the probability that given the state $(X_1, \ldots, X_n)$ at time t, the next reaction will occur in the interval $(t + \tau, t + \tau + d\tau)$, and will be an R_μ reaction. Observe that $P(\tau, \mu)$ is a joint probability density function of continuous variable τ, $\tau \geq 0$, and discrete variable μ, $1 \leq \mu \leq m$.

Theorem 3.8. If $a_0 = \sum_\nu a_\nu$, then

$$P(\tau, \mu) = a_\mu \exp\{-a_0 \tau\}, \quad 0 \leq \tau < \infty, 1 \leq \mu \leq m . \tag{3.24}$$

Proof. Let $P_0(\tau)$ be the probability that given the state $(X_1, \ldots, X_n)$ at time t, no reaction will occur during the next interval of length τ. Then we have

$$P(\tau, \mu)d\tau = P_0(\tau) \cdot a_\mu d\tau . \tag{3.25}$$

But $1 - \sum_\nu a_\nu d\tau'$ is the probability that in state $(X_1, \ldots, X_n)$, no reaction will occur during the next time interval of length $d\tau'$. Thus

$$P_0(\tau' + d\tau') = P_0(\tau') \cdot [1 - \sum_\nu a_\nu d\tau'] \tag{3.26}$$

and hence

$$P_0(\tau) = \exp\{-\sum_\nu a_\nu \tau\} . \tag{3.27}$$

Substituting this expression for $P_0(\tau)$ into Eq. (3.25) yields the result.   □

The direct reaction method is based on the decomposition of the reaction probability density function $P(\mu, \tau)$. This technique is termed conditioning and leads to the equation

$$P(\tau, \mu) = P_1(\tau)P_2(\mu \mid \tau) . \tag{3.28}$$

Here $P_1(\tau)d\tau$ is the probability that the next reaction will occur in the interval $(t + \tau, t + \tau + d\tau)$, and $P_2(\mu \mid \tau)$ is the probability that the next reaction will be an R_μ reaction, given that the next reaction will occur at time $t + \tau$.

The probability $P_1(\tau)d\tau$ is the sum of the probabilities $P(\tau, \mu)d\tau$ over all μ-values. Thus, in view of Eq. (3.24),

$$P_1(\tau) = a_0 \exp\{-a_0\tau\}, \quad 0 \le \tau < \infty . \tag{3.29}$$

Substituting this into Eq. (3.28) yields the discrete probability function

$$P_2(\mu \mid \tau) = \frac{a_\mu}{a_0}, \quad 1 \le \mu \le m . \tag{3.30}$$

As $P_2(\mu \mid \tau)$ is independent of τ, we put $P_2(\mu) = P_2(\mu \mid \tau)$.

The direct reaction algorithm belongs to the class of Monte Carlo methods and simulates the stochastic process described by the probability density function $P(\tau, \mu)$. For this, the stochastic process proceeds in discrete time steps choosing in the actual time instant t a pair (τ, μ) according to density $P(\tau, \mu)$ so that the reaction R_μ occurs at time instant $t + \tau$, and $t + \tau$ becomes the actual time instant. But in view of Eq. (3.28) and the remark in the last paragraph, $P(\tau, \mu)$ can be viewed as a joint probability density function

$$P(\mu, \tau) = P_1(\tau)P_2(\mu) . \tag{3.31}$$

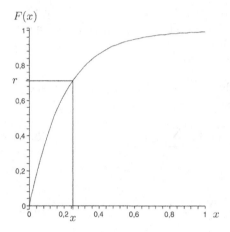

Fig. 3.8 Inversion method.

Thus, a pair (τ, μ) can be drawn so that τ is chosen according to probability density $P_1(\tau)$ and μ is taken according to probability function $P_2(\mu)$. To this end, the *inversion method* (Fig. 3.8) is employed:

In order to generate a random value x according to a given probability density function $P(x)$, draw a random number r from the uniform distribution in the unit interval $[0, 1]$ so that $F(x) = r$, where $F(x) = \int_{-\infty}^{x} P(u)du$ is the corresponding probability distribution function.

First, the distribution function of the density $P_1(\tau)$ is given by

$$F_1(\tau) = \int_{-\infty}^{\tau} P_1(\tau')d\tau' = 1 - \exp\{-a_0\tau\} . \tag{3.32}$$

Take a random number r_1 from the uniformly distributed unit interval and put $F_1(\tau) = r_1$. Resolving for τ (and replacing the random variable $1 - r_1$ by the statistically equivalent random variable r_1) yields

$$\tau = (1/a_0) \log(1/r_1) . \tag{3.33}$$

Second, the (discrete) distribution function of the probability function $P_2(\mu)$ is given as follows:

$$F_2(\mu) = \sum_{\nu=-\infty}^{\mu} P_2(\nu) = \sum_{\nu=1}^{\mu} a_\nu/a_0 . \tag{3.34}$$

Draw a random number r_2 from the uniformly distributed unit interval and take for μ that value which satisfies

$$F_2(\mu - 1) < r_2 \leq F_2(\mu) . \tag{3.35}$$

That means, take μ to be that integer for which

$$\sum_{\nu=1}^{\mu-1} a_\nu < r_2 a_0 \leq \sum_{\nu=1}^{\mu} a_\nu . \tag{3.36}$$

These observations lead to Gillespie's direct reaction algorithm 3.1 (1977).

Example 3.9. Consider the DNA hybridization reaction

$$\text{ssDNA}_1 + \text{ssDNA}_2 \xrightarrow{k} \text{dsDNA} .$$

Algorithm 3.1 GILLESPIE'S DIRECT REACTION METHOD

Input: Stochastic reaction constants $c_1 \ldots , c_m$, initial molecular population numbers $X_1, \ldots , X_n$

1: $t \leftarrow 0$
2: **for** $i \leftarrow 1$ to N **do**
3:     Calculate the propensities $a_\nu = h_\nu c_\nu$, $1 \leq \nu \leq m$.
4:     Generate random numbers r_1 and r_2 from the uniformly distributed unit interval.
5:     Calculate τ and μ according to (3.33) and (3.36), respectively.
6:     $t \leftarrow t + \tau$
7:     Adjust the molecular population levels to reflect the R_μ reaction.
8: **end for**

Take 100,000 oligonucleotides of length 23 nt in a volume V of 10^{-15} l. This gives an approximate concentration of 1.66×10^{-4} mol/l. The melting temperature T_m was set to 338.15 K and the $[\text{Na}^+]$ value was taken to be 4.0 mol/l. The simulation of the reaction by Gillespie's direct reaction method shows that the formation of double-stranded DNA is favored (Fig. 3.9).   ◇

Gillespie's First Reaction Method

Gillespie's algorithm is direct in the sense that it calculates the quantities τ and μ directly. D. Gillespie developed another simulation algorithm which generates a putative time τ_μ for each reaction R_μ to occur and lets μ be the reaction whose putative time comes first and lets τ be the putative time τ_μ. To this end, the putative times τ_μ are drawn according to Eq. (3.33),

$$\tau_\mu = (1/a_0) \log(1/r_\mu) , \tag{3.37}$$

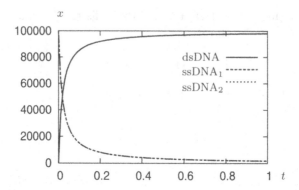

Fig. 3.9 Results of stochastic simulation of DNA hybridization reaction (time (h) vs. number of molecules).

where r_μ is a random number from the uniformly distributed unit interval, $1 \leq \mu \leq m$. These observations lead to Gillespie's first reaction algorithm 3.2 (1976). Both algorithms are equivalent in the sense that the probability distributions used to choose the pair (μ, τ) are the same. A more efficient version of Gillespie's first reaction algorithm is the Gibson-Bruck algorithm (2000).

Algorithm 3.2 GILLESPIE'S FIRST REACTION METHOD

Input: Stochastic reaction constants $c_1 \ldots, c_m$, initial molecular population numbers
    $X_1, \ldots, X_n$

1: $t \leftarrow 0$
2: **for** $i \leftarrow 1$ to N **do**
3:     Calculate the propensities $a_\nu = h_\nu c_\nu$, $1 \leq \nu \leq m$.
4:     **for** $\nu \leftarrow 1$ to m **do**
5:         Generate putative time τ_ν according to (3.37).
6:     **end for**
7:     Let R_μ be the reaction whose putative time τ_μ is smallest.
8:     $t \leftarrow t + \tau_\mu$
9:     Adjust the molecular population levels to reflect the R_μ reaction.
10: **end for**

3.3 Genes

The gene is considered to be the basic unit of inheritance. The genetic material carries the information that directs all physiological activities of the cell and specifies the developmental changes of the multicellular organisms. Understanding the gene structure and function is therefore of fundamental importance.

3.3.1 Structure and Biosynthesis

The stored information in the DNA sequence is expressed in a two-stage process, comprising *transcription* into RNA and *translation* of the nucleotide sequence into an amino acid sequence. The RNA is another type of nucleic acid that differs from the DNA in its sugar (ribose instead of deoxyribose), and the thymine (T) base is replaced by uracil (U). RNA is usually single-stranded and can have different functions in the cell, for example, transfer RNA (tRNA, transfers amino acid to the polypeptide chain) or ribosomal RNA (rRNA, one of the components of the ribosomes). The RNA that transfers the information from the DNA to the ribosomes, which represent the protein synthesis machinery of the cell, is called *messenger RNA* (mRNA). One of the DNA strands, the *template strand*, directs the synthesis of the mRNA via complementary base-pairing by the addition of nucleotides to the 3' end to the growing mRNA (5' to 3' direction of synthesis). The process is controlled by some functional sequence regions that serve as a set of rules to govern the transcription (Fig. 3.10). The transcription is initiated from a promoter located upstream from the DNA coding sequence. Promoter sequences are highly conserved and are recognized by transcription factors, which recruit further the *RNA polymerase*, the enzyme that is responsible for transcribing the coding part of the DNA into mRNA. Different proteins that participate in the transcription process bind to a specific sequence in the double-stranded DNA and usually make contacts with the bases in the major groove of the DNA helix. During elongation, the RNA polymerase moves along the DNA and elongates the RNA, whereas sequential unwinding of the DNA helix precedes the transcription into RNA. Behind the RNA polymerase, the unwound regions base-pair again and restore the original DNA double helix. The process is terminated by a *terminator sequence* located downstream from the coding sequence. Transcription is a fast process: 20 to 50

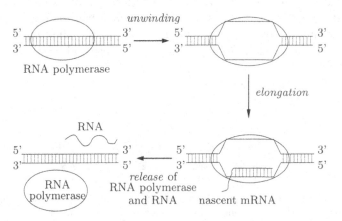

Fig. 3.10 Transcription of DNA into mRNA by RNA polymerase.

nucleotides per second are added to the growing nascent mRNA chain in vivo at 37°C.

In many cases, the function of the single stranded RNA requires a well-defined three-dimensional structure, which also involves base-pairing. In the double-stranded DNA molecules, each base is fixed to its corresponding partner to form a complementary base pair. For the single-stranded RNA (valid also for the single-stranded DNA), each base can bind only one partner at a time, but there are multiple potential partners for such pairing within one chain (Fig. 3.11). With different pairing partners along the chain, various intramolecular, partially double-stranded structures can be formed; each of them is stabilized by the effective free energy of the double-stranded complementary region and the type and the length of the loop enclosed in this structure. The complementary structures formed are imperfect and the integrity can be interrupted by several non-complementary regions, forming three types of loops:

- Hairpin loops: A single chain flips back through a non-complementary region and forms a double strand with adjacent sequences.
- Internal loops: Short regions within a long double-stranded region are not complementary.
- Bulge loops: One of the strands contains bases which are not complementary to the opposite sequence.

The formation of complementary stretches releases free energy (negative value of the Gibbs free energy), which accounts for stabilization, whereas the loops introduce a positive value of free energy (i.e., require energy to be formed). Additional energy is released through the hydrophobic interactions between the base pairs, which are stacked over each other within the double-stranded region. The overall stability of the secondary structure of the single-stranded polynucleotide molecule is determined by the sum of the stabilization through base-pairing and destabilization by the loop structures. The free energy has to reach a sufficient negative value overall, or the secondary structure will be not formed. Intramolecular base-pairing is implicated in the termination of the transcription: a hairpin of palindromic sequences at the 3' terminus causes RNA polymerase to pause and terminate the transcription.

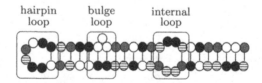

Fig. 3.11 Various intermolecular loops formed by intramolecular base-pairing within a single-stranded RNA molecule. Symbols: open circle = A, black circle = U, gray circle = C, and shaded circle = G.

A *gene* is determined by a segment of the DNA comprising the information necessary to be transcribed into functional RNA and further translated into an amino acid sequence with the flanking regulatory and controlling elements that ensure the fidelity of these processes. The information in the mRNA is read from the 5' to 3' direction in groups of three nucleotides and each trinucleotide or triplet is called a *codon*. The starting point of the translation determines the sequence of the non-overlapping codons, which provides the *reading frame*. A mutation that changes the triplet frame by insertion or deletion of base pairs will cause a change in the reading frame, known as *frameshift*. The new reading frame will generate a completely new RNA sequence beyond the site of mutation. Adverse environmental conditions or errors during replication are sources of mutations. Mutations are rare stochastic events and any base pair can be mutated. A change of only a single base pair is called *point mutation*. The average spontaneous mutation rate corresponds to changes at individual nucleotides of 10^{-9} to 10^{-10} per generation. Substitution mutations without any apparent effect on the amino acid level are designated as *silent mutations*.

In prokaryotes and in some nematodes the same regulatory sequences can govern the transcription of more than one structural gene (i.e., translation into more than one protein). This functional unit of the DNA sequence that comprises many structural genes controlled from a common promoter and transcribed into one mRNA is called *operon*. The activity of the structural genes within an operon is regulated by an operator, a sequence located downstream from the promoter and upstream from the initial AUG codon that determines the start of the mRNA. The operator interacts with the repressor and activator proteins which regulate the transcription of the structural genes and can be encoded within the operon or elsewhere in the genome.

While the prokaryotic genes are colinear with the proteins (i.e., the DNA sequence corresponds exactly to the amino acid sequence), the coding sequence of eukaryotic genes is interrupted by additional non-coding sequences. The coding segments are called *exons* and the non-coding *introns*. The introns are excised from the mRNA and the exons are joined through a process known as *gene splicing* that occurs in the nucleus of the eukaryotic cell. Some mRNAs undergo a self-splicing, but the splicing of the majority of the mRNA is catalyzed by the *spliceosome*, a large RNA-protein complex composed of small nuclear ribonucleo-proteins. Specific recognition signals are located within the intron (e.g., an invariant GU at the intron's 5' boundary and invariant AG at its 3' boundary define the splice junction). Within the introns, around 20 to 50 residues upstream from the 3' splice site is found a conserved sequence in all the vertebrate mRNAs: CURAY, where R represents purines (A or G) and Y represents pyrimidines (C or U). This sequence represents the *branch point*. The splicing is initiated by release of the 5' intron end and joining it to the branch point thereby forming a loop structure, a lariat intermediate. On the second stage, the 3' hydroxyl group of the exon performs a nucleophilic attack at the 3' splice site and ligation of the exons.

The splicing process is not uniform and the exons may be arranged in an alternative pattern, which is known as *alternative splicing*. This process occurs in the higher eukaryotes and is an important mechanism in tissue differentation and developmental regulation of gene expression. Selection of different splicing sites is regulated by serine/arginine residue proteins or SR proteins. Differential use of promoters or termination sites can produce alternative N-termini or C-termini, respectively, in proteins (Fig. 3.12). Alternative reshuffling of the exons or retaining of some of the introns leads to variations in the codons or to a new amino acid sequence. Alternative splicing is an efficient way to economically store larger amounts of genetic information on shorter DNA sequences.

3.3.2 DNA Recombination

DNA recombination refers to a process by which a DNA segment from one DNA molecule is exchanged with a DNA segment from another. A common type of recombination is homologous recombination, known also as *DNA crossover* (Fig. 3.13). When two homologous parts of the DNA align side by

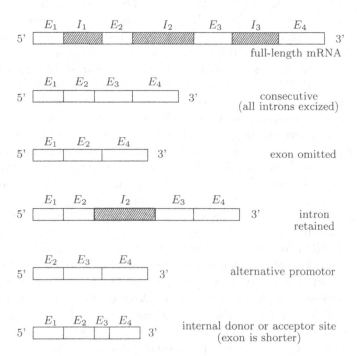

Fig. 3.12 Splicing of mRNA. Introns (I_i) are shaded and exons (E_i) are presented as open structures.

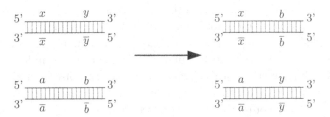

Fig. 3.13 DNA crossover.

side, they can exchange identical parts. In the regions with largely repetitive segments, the alignment might be only partial, which will lead to an unequal crossover. This type of recombination involves the exchange between two homologous DNA molecules, without altering the overall genetic information. The exchange of the strands between homologous DNAs that form heteroduplexes is promoted and guided by specific cellular enzymes. In contrast to the general homologous recombination which necessitates extensive stretches of sequence homology, a site-specific recombination can occur between two sequences with only a small core of homology. The proteins that mediate this process recognize specific target sequences within the DNA, unlike the complementary base-pairing by the homologous recombination. The site-specific recombination creates diversity, thus increasing the array of the proteins that can be synthesized from a certain pool of DNA information. In addition to increasing the genetic diversity, recombination in general plays an essential role for repairing damaged DNA.

3.3.3 Genomes

The complete information carried by the DNA (coding and non-coding sequences) in one organism is referred to as *genome*. In the eukaryotic systems the term genome is specifically applied to the DNA encoded in the nucleus, but the term can also be used for some organelles that contain their own DNA, such as mitochondrial or chloroplast genomes. The total amount of the DNA in one genome is a characteristic feature of each organism and is known as *C-value*, which is defined as the length of the genome. The genomes vary from 10^3 base pairs for some DNA viruses to 10^{11} base pairs for some plants and amphibians. The size of the human genome, for example, is 3×10^9 bp. Large variations in the C-value between similar species are observed: For the amphibious species, the smallest genome is 10^3 bp while the largest is 10^{11} bp. The genome size increases with the rise in complexity of the organisms from prokaryotes to eukaryotes, although in the higher eukaryotes, the proportional correlation between genome size and organism complexity disappears.

The prokaryotic *Escherichia coli* genome consists of 4.7 million base pairs and codes for about 3,000 genes, whereas the 2.9 billion base-pairs haploid human genome is estimated to code for approximately 40,000 genes. The genomes of some lungfishes, however, are larger than those of mammals. Variations in the genome size and in the C-value do not bear the complexity of the organism which gave rise to the C-value paradox. The puzzle of the disproportional C-value to the organism's complexity has been partially solved after identification of non-coding DNA and repetitive DNA sequences.

3.4 Gene Expression

Proteins are the main active players in the biochemical and physiological processes of the cell and implement the unique information that is stored in the ribonucleotid sequences. This task is executed with high fidelity and many steps of control assure the high accuracy of the process.

3.4.1 Protein Biosynthesis

The information in the mRNA is processed in a sequential manner in the 5' to 3' direction, where subsequent codons are translated into amino acids. Each set of three nucleotides corresponds to a specific amino acid, and this genetic code is nearly universal for all living organisms. The four nucleotides (A, U, C, and G) at each of the three positions of a codon form 64 possible codons that encode for only 20 standard amino acids and three non-transcribed *nonsense* or *stop-codons* (UAG, UAA, UGA). Hence, the genetic code is redundant and highly degenerate, and multiple codons encode the same amino acid. Six codons exist for the amino acids arginine (CGU, CGC, CGA, CGG, AGA, AGG), leucine (UUA, UUG, CUU, CUC, CUA, CUG), and serine (AGU, AGC, UCU, UCC, UCA, UCG), and the rest of the amino acids are specified by either four, three or two codons. Only two of the amino acids, methionine and tryptophan, are encoded by a single codon (Table 3.3). The set of codons coding for one amino acid differ mostly in the third position (i.e., the mutation at the third position is phenotypically silent), and the degeneracy of the genetic code might be an evolutionarily acquired tolerance to minimize the deleterious effect of point mutations.

The genetic code is widespread but not universal. The genetic code of mitochondria shows some deviations from the "standard" genetic code for some amino acids. In certain proteins, substituted amino acids can be integrated via standard stop codons: UGA can code for selenocysteine and UAG can code for pyrolysine. The standard genetic code has been expanded to allow incorporation of unnatural amino acids, including amino acids modified with fluorophores for specific detection and chemical and photo-chemical reactive

Table 3.3 The genetic code.

1-Letter Code	3-Letter Code	Codon	1-Letter Code	3-Letter Code	Codon
A	Ala	GCA	L	Leu	CUC
		GCC			CUG
		GCG			CUU
		GCU	K	Lys	AAA
R	Arg	AGA			AAG
		AGG	M	Met	AUG
		CGA	F	Phe	UUC
		CGC			UUU
		CGG	P	Pro	CCA
		CGU			CCC
N	Asn	AAC			CCG
		AAU			CCU
D	Asp	GAC	S	Ser	AGC
		GAU			AGU
C	Cys	UGC			UCA
		UGU			UCC
E	Glu	GAA			UCG
		GAG			UCU
Q	Gln	CAA	T	Thr	ACA
		CAG			ACC
G	Gly	GGA			ACG
		GGC			ACU
		GGG	W	Trp	UGG
		GGU	Y	Tyr	UAC
H	His	CAC			UAU
		CAU	V	Val	GUA
I	Ile	AUA			GUC
		AUC			GUG
		AUU			GUU
L	Leu	UUA	Stop	Ochre	UAA
		UUG		Amber	UAG
		CUA		Opal	UGA

groups for crosslinking. These amino acids are site-specifically incorporated into the peptide chain in response to an amber (UAG) stop codon by an amber suppressor tRNA that is aminoacylated with the desired unnatural amino acid. Another approach of introducing unnatural amino acid relies on the read-through of four-base codons via an aminoacyl-tRNA with an engineered anti-codon loop to accommodate four bases.

Translation of the mRNA is carried out by the ribosomes and each codon binds in a complementary manner to the unpaired bases, known as *anti-codon* of the corresponding tRNA (Fig. 3.14). Transfer RNA is a small non-coding RNA chain (60 to 95 nucleotides) with a cloverleaf secondary structure. It is the carrier for the amino acid, which after the base pairing of the codon and anti-codon is covalently linked to the carboxyl terminus of the growing

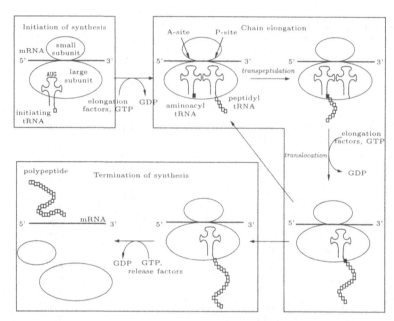

Fig. 3.14 Translation of the genetic information from mRNA to polypeptide chain.

polypeptide chain. The proper tRNA is selected based on the complementary codon-anticodon interactions. Although many organisms contain different isoaccepting tRNAs (different tRNAs specific for the same amino acid), in some cases the same tRNA can bind to two or three codons, encoding the same amino acid. The codon-anticodon pairing in this case is incomplete with a non-Watson-Crick geometry at the third position, known as a *wobble base pair*. Only certain wobble base-pairs are allowed (G-U and I-U, I-A, I-C, where I is a modified base) whose thermodynamic stability is similar to the Watson-Crick base pair.

Ribosomes are large nucleoprotein complexes consisting of two subunits that read the genetic code on the mRNA in the 5' to 3' direction and translate it into amino acids. Initiation of the translation is a complex process, which requires initiation factors that help the small and large subunits of each ribosome to assemble on the mRNA with the first aminoacyl-tRNA (Fig. 3.14). This process is slow and rate-limiting for the translation. The translation starts at the AUG codon, which is recognized by the initiator tRNA for methionine. In eukaryotic cells only initiator tRNA can bind to the small subunit of ribosomes before it assembles with the large subunit on the mRNA. In prokaryotic organisms the recognition of the first AUG-codon is controlled by a specific ribosome-binding nucleotide sequence (6 bp) upstream from the start AUG-codon. Therefore, in bacteria the translation can be initiated at any AUG-position in the mRNA given the presence of the upstream ribosome-binding sequence. As a consequence, the mRNA in prokaryotes is often polycystronic (i.e., one mRNA molecule can encode for many proteins). Ribosomes elongate the polypeptide chain by adding one amino acid residue at the time in a three-stage reaction cycle (Fig. 3.14). The tRNA charged with an amino acid binds to the acceptor site (A site) of the ribosome at the $n + 1$-th codon. The part of the tRNA that carries the amino acid is in the large subunit, whereas the anti-codon at the other end binds to the mRNA codon. The n-th codon that has been read is in the donor site (P site), which is occupied by the peptidyl-tRNA carrying the nascent amino acid chain. A peptide bond is formed in the second stage through a nucleophilic displacement of the peptidyl-tRNA by the 3'-linked amino acid of the aminoacyl-tRNA. The reaction occurs in ATP-independent manner and the energy is provided by the high energy bond between the polypeptide and the peptidyl-tRNA. The new amino acid from the tRNA is transferred to the C-terminus at the growing chain and the *transpeptidation* is catalyzed by the peptidyltranferase activity of the large subunit. In the third and final stage, the new peptidyl-tRNA in the A site is transferred, together with the bound codon of mRNA, to the P site. The efficiency of this translocation process is maintained by an elongation factor that binds to the ribosome together with GTP, delivering the energy for the transfer reaction. The elongation of the polypeptide nascent chain is the most rapid step in the protein synthesis. When the ribosomes encounter the stop codon the synthesis is terminated, which results in a release of the polypeptide and dissociation of the two ribosomal subunits from the mRNA. The termination is facilitated by release factors and GTP-hydrolysis. In the whole biosynthetic cycle of the polypeptide chain, GTP acts as an energy donor, ensuring fastness and irreversibility of the coupled initiation, elongation and termination of the translation.

3.4.2 Proteins – Molecular Structure

Proteins are linear polymers which have to acquire a correct 3D structure to accomplish the physiological functions in the cell. The ribosomes decode the information from the mRNA and translate it into a polypeptide chain, built from the 20 amino acids. The linear sequence of the amino acids that is determined from the genetic information stored in the DNA determines the *primary protein sequence*. The amino acids are linked in a dehydration reaction (release of water) by the C-terminal COOH group of the n-th amino acid and the N-terminal NH$_2$ group of the $n + 1$-th amino acid forming a covalent peptide bond $(-CO-NH-)$. The backbone of each protein is identical, while the side chains of the different amino acids introduce diversity into the physicochemical properties of each protein. The differences in the chemical structure of the individual amino acids determines the directionality of the protein chain; the peptide chain starts with a free amino group (N-terminus) of the first amino acid and ends with a free carboxyl group of the last amino acid (C-terminus).

Short or large sequence regions are forming secondary structures stabilized by non-covalent interactions (i.e., hydrogen bonds, hydrophobic interactions) (Fig. 3.15). The most common structures are helices and beta-sheets. Helices are stabilized with hydrogen bonds between the C = O group of the peptide bond of the n-th residue with the NH group of the $n + 4$-th amino acid. In addition, the tight packing of the helix allows van der Waals interactions of the atoms across the helix. The most common α-helix is right handed and can accommodate on average 3.6 residues per turn. Helices in the proteins range from four to over forty residues, but a typical helix spans about 12 residues, corresponding to over three helical turns.

The second structural element is the beta pleated sheet consisting of strands connected with a hydrogen bond network between NH groups in the backbone of one strand and C = O groups of the adjacent strand, forming a twisted pleated sheet. The backbone hydrogen bonding of the beta sheets is generally considered as slightly stronger than that found in the α-helices. The two neighboring chains can run in the same (parallel beta pleated sheet) or in the opposite direction (anti-parallel beta pleated sheet). Helices and beta-sheets comprise around half of the local structures in the globular proteins. The remaining parts have either coil or loop conformation. Turns and loops establish the joints between different secondary structural elements and almost always occur at the surface of the protein. Turns are also stabilized by a hydrogen bond, either: between the n-th and $n + 3$-th residues (β-turn); between the n-th and $n + 2$-th residues (γ-turn); between the n-th and $n + 4$-th residues (α-turn); between the n-th and $n + 5$-th residues

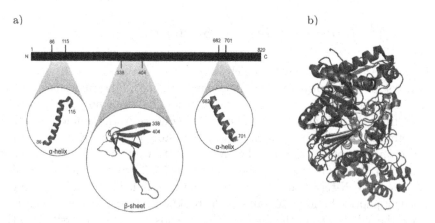

Fig. 3.15 Protein structure exemplified by pro-penicillin amidase (protein data bank, PDB, code 1E3A): **a** The primary amino acid amino acid sequence can form different secondary structures; **b** The backbone of the 3D structure. The crystal structure of pro-penicillin amidase was solved by L. Hewitt and coworkers (2000).

(π-turn). The Ω-loop is a term for a longer loop (named because of its similarity to the Greek upper case letter omega) comprising 6 to 16 residues with an end-to-end distance of less than 10 Å. The Ω-loop also adopts a compact conformation, in which the side-chains tend to fill the internal loop cavity.

The secondary structures are spatially ordered to one another in the 3D space forming the *tertiary structure*. Non-polar residues (e.g., Leu, Ile, Val, Phe) are sequestered in the interior of the molecule in the globular proteins, thus avoiding the contact with water from the environment. In turn, the proteins' surfaces are enriched with charged polar residues (e.g., Arg, Lys, His, Asp, Glu) that can establish contacts with the aqueous exterior. In special cases, these residues can be found in the interior of the protein, where they accomplish specific functions (e.g., catalytic functions in the enzymes). Uncharged polar residues (e.g., Ser, Asn, Gln, Tyr) can be found both on the surface and in the interior of the molecule. When found buried in the interior, they are almost always involved in hydrogen bonds. Tertiary structure (i.e., the long- or short-range interactions between the side chain of the amino acids), and the propensity to establish different secondary structures, are assumed to be determined by the primary sequence of the protein. Various secondary interactions between the backbone and different side chains provide the structural basis for the native 3D pattern of a protein sequence: (1) electrostatic interactions, including dipole-dipole and ionic interactions, (2) hydrogen bonds, (3) hydrophobic interactions, and (4) covalent disulfide bonds, formed between some free sulfhydryl groups in the side chains of the cysteines in an oxidizing environment. The tertiary structural arrangement of the proteins ensures dense packing with a minimized ratio of the volume enclosed in van der Waals interactions (Fig. 3.15). This ratio of about 0.75 is within the same range of crystals formed from small organic molecules.

The 3D structure, in which a protein can accomplish its physiological functions in the cell, is also called native fold or native conformation. It is commonly assumed that in the cellular environment, the native conformation is the most thermodynamically stable conformation, populating the global minimum of the energy. En route to the stable native fold, some proteins go through several partially folded intermediate states, which dwell in local energy minima. In the cell, a variety of other proteins (e.g., chaperones, disulfide isomerases, catalyzing the formation of disulfide bonds) assist the newly synthesized proteins in attaining their native conformation. For some proteins, the dwelling in a local energy minimum might have a physiological role. A classical example is the hemagglutinin, in which the two chains of the mature protein are kinetically trapped in an intermediate state. A drop in the pH causes conformational changes in the intermediate to an energetically more favorable state, which enables it to penetrate the host cell membrane. Proteins are not rigid molecules and might shuttle between different structures or conformations, which allows them to accomplish their physiological functions (e.g., binding of substrates to the active site of enzymes).

Several polypeptide chains that have independently acquired a 3D structure can interact in order to attain a highly ordered structure, a *quaternary structure*. The individual chains are called subunits and they are stabilized by the same type of interactions, responsible for the stability of the 3D structure (i.e., secondary non-covalent interactions or disulfide bonds within the protein complex). In one such complex two (dimer), three (trimer) or more polypeptides (multimer) can be associated. Proteins consisting of identical subunits only are referred to with a prefix "homo-" (homodimer, homotrimer, etc.), whereas the complexes containing structurally different subunits are assigned the prefix "hetero-" (heterodimer, heterotrimer, etc.).

3.4.3 Enzymes

Enzymes are proteins that accomplish specific catalytic functions in the cell. They enormously accelerate chemical reactions in the cell (10^6–10^{12} increase of the rate over the corresponding uncatalyzed reaction). Enzymes are specific and act as catalysts only in one or a few similar reactions and are involved in all biochemical reactions in the cell (e.g., DNA replication, RNA transcription, catabolic (degradation) and anabolic (synthesis) reactions). A small fraction of the amino acids of the entire enzyme molecule form the catalytic, active center that comes into contact with the *substrate*, a molecule that will be converted to one or more *products*. This segment, usually consisting of three or four residues, is called the *active site*. Similar to the catalytic mechanism of the chemical catalysts, the enzymes lower the activation energy of the reaction (E_a or $\Delta G^{\ddagger}$), thus accelerating the rate of reaction; they do not change the reaction pathway and do not alter the equilibrium (Fig. 3.16). Unlike the chemical catalysts the enzymes rarely produce side products and show a significant level of stereospecificity (recognize only one stereoisomer as a substrate) and regioselectivity (specific for only one substrate or small range of related compounds). The most specific enzymes are involved in the amplification and storage of the genome information (e.g., DNA polymerase, RNA polymerase, ribosomes, aminoacyl-tRNA synthase). Mammalian polymerases have an extreme fidelity with an error rate of about 10^{-7}. Some enzymes in the secondary metabolic pathways are described as promiscuous, because they can act on a broad range of different (but structurally similar) substrates.

The substrate binds to the enzyme through geometrically and physically complementary interactions and the binding is controlled through non-covalent forces which are identical with the interactions stabilizing different protein conformations. The first theory of the enzyme-substrate

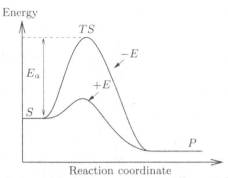

Fig. 3.16 Energy landscape of enzyme-catalyzed $(+E)$ and uncatalyzed $(-E)$ conversion of substrate (S) into product (P). When processing from higher energy S to lower energy P, the substrate must first pass the transition state TS, and the energy to reach the activation energy E_a represents the energy barrier of the reaction.

binding was developed by E. Fischer in 1894 suggesting that both molecules fit to each other based on a geometrically similar shape (e.g., "the lock and key" model). Although specificity is well covered by this model, it cannot explain the catalytic mechanism. Koshland's theory (1958), known as *induced fit*, suggests that substrate and enzyme are both flexible molecules and interactions between them can reshape their conformational state.

Some enzymes require other factors, *cofactors*, to be bound to accomplish their activity. Cofactors can be either inorganic (metal ions) or small organic molecules (heme, flavin, NADH, vitamins). Cofactors that remain tightly bound to the enzyme, whose function they assist, are known as *prosthetic groups* (NADH). Enzymes with a bound cofactor are called holoenzymes (i.e., active form), and enzymes without a cofactor apoenzymes. Co-enzymes are chemically changed during the enzyme reaction and regenerated at the end, thus maintaining constant steady-state levels inside the cell.

Enzyme activity can also be affected by other molecules, called *effectors*, which modulate the enzyme activity either by direct binding to binding sites of the enzyme or indirectly through subunits that modulate the enzyme function. Such enzymes are called *allosteric*. Small molecules can also decrease or inhibit enzyme activity (i.e., inhibitors), or enhance the enzyme activity (i.e., activators). Enzymes, like all the proteins, are conformationally dynamic and this internal dynamics can be essential for the catalytic properties. Temperature and environmental pH can influence the conformational freedom of the enzyme molecule and change their catalytic properties. In the cell, the enzyme activity is controlled additionally by its amount, which depends on the rate of biosynthesis and the rate of degradation.

Enzyme Kinetics

The conversion of a substrate S to a product P catalyzed by an enzyme E can be described by the chemical equation

$$E + S \underset{k_{-1}}{\overset{k_1}{\rightleftharpoons}} ES \overset{k_2}{\longrightarrow} E + P . \tag{3.38}$$

This process undergoes two stages. First, the substrate reversibly binds to the enzyme and an enzyme-substrate complex ES is formed. Second, the enzyme catalyzes the chemical step in the reaction and releases a product. The enzyme is not altered by the reaction and thus the equilibrium is influenced only by the thermodynamical properties of S and P. In equilibrium, the conversion to the product becomes rate-limiting. Therefore, the corresponding differential reaction-rate equation has the property

$$\frac{d[ES]}{dt} = k_1[E][S] - k_{-1}[ES] - k_2[ES] = 0 . \tag{3.39}$$

Hence,

$$[ES] = \frac{k_1[E][S]}{k_{-1} + k_2} . \tag{3.40}$$

In 1913, L. Michaelis and M. Menten assumed that product formation is the rate-limiting step ($k_{-1} \gg k_2$). The maintenance of the ES complex in equilibrium (steady-state assumption) is then described by the *Michaelis constant*

$$K_m = \frac{k_{-1} + k_2}{k_1} \approx \frac{k_{-1}}{k_1} . \tag{3.41}$$

The non-covalent ES complex is known as the Michaelis complex. Eq. (3.40) then simplifies to

$$[ES] = \frac{[E][S]}{K_m} . \tag{3.42}$$

The total concentration of the enzyme is the sum of the concentrations of the enzyme in the ES complex and the free soluble enzyme,

$$[E_0] = [E] + [ES] . \tag{3.43}$$

Thus, Eq. (3.42) rearranges to

$$[ES] = \frac{([E_0] - [ES])[S]}{K_m} , \tag{3.44}$$

which can further be transformed to

$$[ES] = [E_0] \frac{1}{1 + K_m/[S]} . \tag{3.45}$$

The initial velocity of the reaction is given by

$$\frac{d[P]}{dt} = k_2[ES] \ . \tag{3.46}$$

Substituting Eq. (3.45) into Eq. (3.46) and multiplying both nominator and denominator by $[S]$ gives

$$\frac{d[P]}{dt} = k_2[E_0]\frac{[S]}{K_m + [S]} = V_{max}\frac{[S]}{K_m + [S]} \tag{3.47}$$

where V_{max} is the maximum velocity. In this case, all enzyme molecules are saturated by the substrate and the enzyme exists only in an ES form:

$$V_{max} = k_2[E_0] \ . \tag{3.48}$$

k_2 is called *catalytic constant* or *turnover number* k_{cat} and gives the number of conversions (turnovers) that each catalytic site catalyzes per unit of time. Combining Eqs. (3.47) and (3.48), an equation known as the *Michaelis-Menten equation* is derived,

$$\nu = \frac{V_{max}[S]}{K_m + [S]} \ . \tag{3.49}$$

At the substrate concentration $[S] = K_m$, the reaction velocity is half-maximal. The Michaelis constant K_m is the measure of the affinity of the enzyme to the substrate: For enzymes with small K_m values, the maximal catalytic activity is achieved at low substrate concentrations. Each enzyme has a characteristic K_m value for a given substrate; the K_m value varies for different substrates of the same enzyme and it is a function of the temperature and pH value. The apparent second-order rate constant k_{cat}/K_m, known also as a *specificity constant*, is used as a measure of the catalytic efficiency of the enzyme. It depends on the encounters between substrate and enzyme in a solution and summarizes both characteristics of an enzyme: affinity and catalytic ability. The specificity constant can be used for the comparison of enzymes with different substrates. The theoretical maximum of k_{cat}/K_m is 10^8–10^9 per Ms and is called diffusion limit (i.e., the diffusion rate limits the reaction rate and every collision of the enzyme with a substrate will release a product).

Michaelis-Menten kinetics assumes irreversibility of the enzymatic catalysis and it is based on the law of mass action, assuming free diffusion and random collision. In the cell, however, the processes can deviate from such idealized conditions. The highly crowded internal space of the cell significantly limits the free molecular movements (e.g., the concentration of macromolecules in the cytoplasm of prokaryotic cells is about 400 mg/ml). For some heterogeneous processes (e.g., as in the case of DNA polymerase), the substrate mobility may also be limited. These deviations from the conventional mass-action laws led to the development of the limited-mobility derived or fractal-like kinetics, reviewed by R. Kopelman (1988), M. Savageau (1995), and S. Schnell and T. Turner (2004).

Enzyme Thermodynamics

The enzymatic reaction (3.38) can be described by the equilibrium equation

$$-RT \log K_{eq} = \Delta G^\circ_{eq}, \tag{3.50}$$

where K_{eq} is the equilibrium constant, ΔG°_{eq} is the Gibbs free energy of the ES complex, T is the absolute temperature, and R is the gas constant. But if it is assumed that the formation of the ES complex is in rapid equilibrium with the reactants, then the equilibrium constant can be expressed as

$$K_{eq} = \frac{[ES]}{[E]\,[S]}. \tag{3.51}$$

Combining Eqs. (3.46), (3.50), and (3.51) gives

$$\frac{d[P]}{dt} = k_2[E][S]e^{-\Delta G^\circ_{eq}/RT}. \tag{3.52}$$

This equation shows that the rate of the reaction depends not only on the reactants, but increases exponentially with ΔG°_{eq}.

3.5 Cells and Organisms

Despite the phenotypic and genotypic variation of different organisms, they share remarkable similarities on the cellular level (e.g., in the general cell structure, physiology and biochemistry). The cells propagate by duplicating the DNA and each daughter cell inherits identical genetic material from the parental cell. This genetic information is translated in a variety of proteins that determine the functional diversity of each cell. The prokaryotic cells are commonly unicellular and a single cell is capable of executing all the physiological activities. Unlike the prokaryotes, in the eukaryotic cell the physiological activities are spatially separated in compartments or organelles surrounded by double-lipid-layer membranes. The various membrane-separated organelles in the cell that accomplish different functions are referred to as the *endomembrane system*. Another level of complexity exists in multicellular eukaryotic organisms: various physiological activities are separated in different cells and have led to cell specialization and differentiation of the cell types.

3.5.1 Eukaryotes and Prokaryotes

The nucleus is the largest organelle in the eukaryotic cell and carries the inheritance information encoded in the DNA. It is surrounded by a double membrane, referred to as the *nuclear envelope*, with pores that allow nucleus components to move in and out. The nuclear membrane extends in various tube- and sheet-like membrane extensions, forming an extensive network of membranes, called the *endoplasmic reticulum*. In eukaryotes, the mRNA is synthesized (transcription) in the nucleus, whereas the translation into proteins occurs in the cytosol. Ribosomes that perform the protein synthesis are attached to the endoplasmic reticulum. Furthermore, the newly synthesized proteins are modified and transported to their final destinations in the *Golgi-bodies* which bud off from the endoplasmic reticulum. The endoplasmic reticulum and Golgi apparatus function not only in the processing and transport of proteins designed for secretion and membrane incorporation, but also actively participate in the synthesis of lipids. Different vesicles can be formed by budding off the membranes whose function is to transport nutrients into and waste products out of the cell.

In nearly all eukaryotes the aerobic respiratory functions are accomplished in the mitochondria, which are surrounded by a double layer of membranes. Mitochondria are found in almost all eukaryotes and play a critical role in energy metabolism. They produce the energy substance, ATP, by oxidative breakdown of the nutrients. Mitochondria contain their own DNA. There is a variety of other simple compartments surrounded by membranes that are responsible for different functions found in the eukaryotic cell:

- Lysosomes contain proteolytic enzymes that digest the food substances.
- Peroxisomes allow specific reactions in a defined environment to take place in which toxic peroxide is released.
- Vacuoles maintain the osmotic pressure of the higher plants.
- Chloroplasts produce energy through photosynthesis and are characteristic of plant cells and various groups of algae.

Next to the spatial separation of biochemical processes, the eukaryotic cell has another level of organization: a network of filamentous proteins throughout the cytoplasm forms the *cytoskeleton*, which provides the cell cytoplasm with structure and shape. The cytoskeleton determines the general organization within the cell and enables motions of the entire cell and throughout the cytoplasm. The cytoskeleton is composed mainly of actin filaments, intermediate filaments, and microtubules.

Prokaryotes are referred to as organisms whose genetic information is not stored in the nucleus or any membrane-bound structure. They are usually smaller and their cell structure is simpler than that of eukaryotes. The genetic material is encoded by a single molecule of chromosomal DNA called also *nucleoid*. Some prokaryotic cells might contain self-replicating satellite circular DNA, known as *plasmids*, which encodes some crucial functions for the prokaryotic cell. The prokaryotes also lack several membrane-bound cell compartments; nevertheless, the biochemical processes are spatially separated within the cell, positioned at differents subcellular sites. Cytoskeleton elements, homologous to eukaryotic cytoskeletal proteins, have also been found in prokaryotes and they function in actively positioning proteins and DNA molecules. Prokaryotes have a shorter generation time and a large surface-to-volume ratio that consequently gives them a higher metabolic rate compared to eukaryotes. Based on the ability of prokaryotes to utilize oxygen or different compounds for oxidation processes, the cells are divided into aerob (use oxygen) and anaerob.

3.6 Viruses

Viruses are sub-microscopic particles that can infect the cells of living organisms. They are not considered as eukaryotes or prokaryotes.

3.6.1 General Structure and Classification

Viruses (translated from Latin as toxin or poison), also known as *virions*, infect both eukaryotic and prokaryotic cells and replicate and propagate further using the transcription and translation systems of the host cell. The group of viruses infecting bacteria is known as bacteriophages. The majority of the viruses have a size of 10–250 nm and with some exceptions up to 750 nm, which is larger than the size of some bacteria. The genetic information is stored on nucleic acid, which can be either RNA (single- or double-stranded) or DNA (single- or double-stranded), and is encapsulated by a protein shell, known as *capsid*. The proteins of the capsid are encoded by the viral genome. Viral particles have regular shapes including helical capsids (tobacco mosaic virus), icosahedral symmetry (hollow quasi-spherical structure; polio virus, and foot and mouth disease virus), enveloped viruses (in addition to the capsid, they are covered by a lipid-bilayer membrane; influenza virus and human immunodeficiency virus), and complex viruses (tailed bacteriophages, poxviruses). Some viruses are unable to survive outside the host cell, whereas others are more stable and can persist outside a cellular environment for very long periods.

According to the classification proposed by the Nobel Prize-winner D. Baltimore, also known as Baltimore classification, viruses can be divided into seven groups: (I) double-stranded DNA, (II) single-stranded DNA, (III) double-stranded RNA, (IV) positive-sense (+) single-stranded RNA, (V) negative-sense (−) single-stranded RNA, (VI) single-stranded-RNA reverse transcribing, and (VII) double-stranded-DNA-reverse transcribing. In the DNA viruses (Groups I and II), the genetic information is stored on DNA and it is replicated via DNA-dependent DNA polymerase, while in the viruses belonging to Groups III, IV, and V the genetic information is stored in an RNA sequence. The ribosomes of the host directly translate the RNA strand into the positive-sense viruses, whereas in the negative-sense viruses the RNA is first inverted into positive-sense RNA via RNA polymerase. Examples of RNA viruses are hepatitis A and C, SARS, yellow fever, rubella and influenza viruses; to the DNA viruses belong simian virus (SV) 40, human herpes virus, cowpox smallpox viruses, and bacteriophages. The most severe Marburg, Ebola, and Lassa viruses belong to the group of negative-sense (−) single-stranded RNA viruses.

In the reverse transcribing single-stranded RNA viruses (Group VI), also called retroviruses, the genetic information is stored on RNA, and they replicate by formation of DNA due to reverse transcription via RNA-dependent DNA polymerase. The DNA is then integrated into the host genome and further propagated by the replication and translation machinery of the host cell. The most prominent member of this group is the human immunodeficiency virus (HIV). The double-stranded DNA of the reverse transcribing double-stranded DNA viruses (Group VII) is transcribed both into mRNA and translated further into protein and RNA, which is integrated into the host genome after reverse transcription into DNA. To the latter Group belongs the hepatitis B virus.

Therapeutical approaches against viral diseases include vaccination or administration of anti-genic material to trigger immunity responses to the disease agent. Antibiotics and other drugs are often applied too, although their targets are mainly inflammation and other secondary responses caused by the viral infection.

3.6.2 Applications

Viruses have a great potential in the virotherapy also called oncolytic or viral therapy. This technique involves specific targeting and killing of cancer cells through the introduction into the body of genetically modified viral material. The viral particles reproduce rapidly over a short period of time and attack only the cancerous cells without harming the healthy cells. In the viruses tailored for the viral therapy, either the protein coat is modified so that the affinity only against cancer cells is assured, or the genetic information of the virus is altered and it can enter any cell but replicate only in the cancerous cells. This therapy can be applied to treat cancer as well as to inhibit angiogenesis. A critical barrier in the widespread application of cancer treatment is the immune system of the host, which responds to the engineered viruses and destroys them.

The shape and size of the viruses, the functional groups on their surface, and the tools they have developed to cross barriers of the host cells make viruses very attractive to material sciences and nanotechnology. The precisely defined pattern of functional groups and their ordering on the virus surface offers a unique scaffold for covalently linked surface modifications. A. Belcher and colleagues (2002) have engineered a liquid crystal system from genetically engineered bacteriophage and zinc sulfide, which spontaneously assembles into a thin hybrid nanocrystal ordered into approximately 72 μm domains. In a subsequent study (2006), they have extended the application of the virus-templated synthesis to assemble nanowires of hybrid gold-cobalt oxide. The virus particles were genetically modified to incorporate gold-binding peptides into the filament coat. The total negative charge of these particles allows them to be layered between oppositely charged polymers to form thin and flexible sheets. This two-dimensional assembly of viruses on polyelectrolyte multilayers is a step forward in creating flexible ion batteries that supply much electrical energy in a thin and lightweight package.

HIV

Human immunodeficiency virus type 1 (HIV-1) is the etiologic agent of the immune deficiency syndrome (AIDS) and its related disorders. The first case report of AIDS appeared in 1981, while the virus was first isolated in 1983. Currently, there are about 49,000 HIV-1 infections in Germany alone, with about 2,900 new infections per year. Primarily, HIV-1 infects and kills immune cells which regulate and amplify immune response. Without effective anti-retroviral therapy, the hallmark decrease in immune cells during AIDS results in a weakened immune system that impairs the ability to fight against infections or cancer, so that death eventually results. HIV-1 is a retrovirus, and similar to all retroviruses its RNA genome replicates by a DNA intermediate. Many retroviruses contain three structural genes (Gag, Pol, and Env, encoding for core and structural proteins, reverse transcriptase, and coat proteins, respectively). HIV-1 has additional accessory and regulatory genes: Vif, Vpr, Vpu (accessory) and Nef, Tat, Rev (regulatory).

The HIV-1 life cycle is well-documented. HIV infection begins when the viral glycoprotein (gp) interacts with the CD4 receptor on the surface of an immune cell. After fusion of the viral membrane with the cell membrane, the viral core with the associated RNA is internalized into the cell. Partial uncoating of the viral core exposes the viral RNA. In the cytoplasm of the recipient cell, the viral genomic RNA is synthesized by viral reverse transcriptase into a viral double-stranded DNA preintegration complex. After migrating into the nucleus, facilitated by other viral proteins the viral double-stranded DNA of the preintegration complex is integrated randomly into the host DNA by viral integrase. RNA polymerase transcribes the proviral DNA into mRNA. In the early transcription phase, the mRNA is spliced by the cellular splicing machinery into multiply spliced transcripts, producing the Tat, Rev and Nef proteins. When Rev accumulates to a critical level, the mRNA production shifts from multiple spliced to single spliced and unspliced transcripts, resulting in the Env gp160 (containing envelope proteins gp120 and gp41), Vif, Vpr, Vpu proteins and Gag p55 (containing matrix, capsid, nucleocapsid) and Gag-Pol p160 (containing matrix, capsid, protease, reverse transcriptase, and integrase) proteins, respectively. The virus is assembled at the plasma membrane, forming a budding virion. After virus budding out from the cell surface into extracellular space, maturation of the virus proceeds (by proteolytical cleavage of the Gag and Gag-Pol polyprotein). Now the virus is ready for another round of infection.

Hepatitis B Virus

Hepatitis B virus (HBV) has one of the smallest genomes (approximately 3 kb) and belongs to the group of reverse transcribing double-stranded DNA viruses. The virus specifically infects the liver of humans and various animals and causes acute liver damage. Over 250 million people worldwide are infected with HBV yearly, some of whom develop severe pathological consequences, including chronic hepatitis, cirrhosis, and hepatocellular cancer. HBV infection is particularly common in Asia and Africa and is associated with approximately 10% of the worldwide liver cancer incidence (up to a million cases of liver cancer annually). Transfection occurs through infected blood and other body fluids.

The hepatitis B virus is composed of an outer lipid envelope and an icosahedral nucleocapsid core (i.e., DNA genome and protein coat surrounding it). The virus attacks the surface receptors of the hepatocytes and after internalization into the cell migrates into the nucleus. It replicates through reverse transcription of RNA intermediate, the pregenomic RNA, which can be packed into capsids and is reversely transcribed into DNA. The new virions bud out of the endoplasmic reticulum and are exported from the cell.

In the cases of acute HBV infection, up to 95% of adults overcome the infection spontaneously without any treatment. Chronic HBV is treated with anti-viral agents that either inhibit the virus replication or stimulate the immune response. Five drugs have been approved for the treatment of chronic HBV infection: interferon-alpha, pegylated interferon-alpha, lamivudine, adefovir dipivoxil, and entecavir. Their efficacy is limited by their side effects, as well the high frequency of viral mutations which render the therapeutics less potent.

References

1. Ambrogelly A, Palioura S, Soll D (2007) Natural expansion of the genetic code. Nat Chem Biol 3:29–35
2. Bock A, Forchhammer K, Heider J, Baron C (1991) Selenoportein synthesis: an expansion of the genetic code. Trends Biochem Sci 16:463–467
3. Breslauer KJ, Frank R, Blöcker H, Marky LA (1986) Proc Natl Acad Sci 83: 3746–3750

4. Brett D, Pospisil H, Valcárcel J, Reich J, Bork P (2001) Alternative splicing and genome complexity. Nature Gen 30: 29–30
5. Bundschuh R, Gerland U (2006) Dynamics of intramolecular recognition: base-pairing is DNA/RNA near and far from equilibrium. Eur Phys J 10:319–329
6. Coffin JM (1996). Retroviridae. In: Fields BN, Knipte DM, Howley PM (eds.) Fields Virology Raven Publ
7. Collier J, Shapiro L (2007) Spatial complexity and control of a bacterial cell cycle. Curr Opin Biotechnol 18:333-340
8. Gibson M, Bruck J (2000) Efficient exact stochastic simulation of chemical systems with many species and many channels. J Phys Chem A 104:1876–1889
9. Gillespie D (1976) A general method for numerically simulating the stochastic time evolution of coupled chemical reactions. J Phys Chem 22:403–434
10. Gillespie D (1977) Exact stochastic simulation of coupled chemical reactions. J Phys Chem 81:2340–2361
11. Hewitt L, Kasche V, Lummer K, Lewist RJ, Murshudov GN, Verma GN (2000) Structure of a slow processing precursor penicillin acylase from Escherichia coli reveals the linker peptide blocking the active cleft. J Mol Biol 302:887–898
12. Hohsaka T, Sisido M (2002). Incorporation of non-natural amino acids into proteins. Curr Opin Chem Biol 6:809–815
13. Jimenez-Sanchez A (1995) On the origin and evolution of the genetic code. J Mol Evol 41:712–716
14. Kelly E, Russell SJ (2007) History of oncolytic viruses: genesis to genetic engineering. Mol Ther 15:651–659
15. Kopelman R (1988) Fractal reaction kinetics. Science 241:1620–1626
16. Lee SW, Mao C, Flynn CE, Belcher AM (2002) Ordering of quantum dots using genetically engineered viruses. Science 296:892–895
17. Mattick JS, Makunin IV (2006) Non-coding RNA, Hum Mol Gen 15:17–29
18. McMahon BJ (2005) Epidemiology and natural history of hepatitis B. Semin Liver Dis 25:3–8
19. Nam KT, Kim DW, Yoo PJ, Chiang CY, Meethong N, Hammond PT, Chiang YM, Belcher AM (2006) Virus-enabled synthesis and assembly of nanowires for lithium ion battery electrodes. Science 312:885–888
20. Noad R, Roy P (2003) Virus-like particles as immunogens. Trends Microbiol 11:438–444
21. Ryu WS (2003) Molecular aspects of hepatitis B viral infection and the viral carcinogenesis. J Biochem Mol Biol 36:138–143
22. SantaLuica J Jr, Hicks D (2004) The thermodynamics of DNA structural motifs. Annu Rev Biomol Struct 33:415–440
23. Savageau MA (1995) Michaelis-Menten mechanism reconsidered: implications of fractal kinetics. J Theor Biol 176:115–24
24. Schnell S, Turner TE (2004) Reaction kinetics in intracellular environments with macromolecular crowding: simulations and rate laws. Prog Biophys Mol Biol 85:235–260
25. Tijssen K (1993) Laboratory techniques in biochemistry and molecular biology. Elsevier, Amsterdam
26. Varani G, McClain W (2000) The G×U wobble base pair. A fundamental building block of RNA structure crucial to RNA function in diverse biological systems. EMBO Rep 1:18–23
27. Vaha-Koskela MJ, Heikkila JE, Hinkkanen AE (2007) Oncolytic viruses in cancer therapy. Cancer Lett 254:178–216
28. Wagner E (2007) Programmed drug delivery: nanosystems for tumor targeting. Expert Opin Biol Ther 7:587–593
29. Wilkinson DJ (2006) Stochastic modelling for systems biology. Chapman & Hall, New York
30. Xie J, Schultz PG (2006) A chemical toolkit for proteins – an expanded genetic code. Nat Rev Mol Cell Bio 7:775–782

Chapter 4
Word Design for DNA Computing

Abstract This chapter addresses the problem of negative word design: Construct a large set of oligonucleotides which selectively hybridize so that undesired molecules encoding false results or blocking the desired reactions are excluded. In practice, such a set of oligonucleotides is designed so that it simultaneously satisfies several thermodynamical and combinatorial constraints.

4.1 Constraints

DNA-based computations rely on short single-stranded DNA molecules referred to as oligonucleotides. The computational process usually allows oligonucleotide strands to hybridize in order to form longer DNA molecules. For this, the oligonucleotide strands called *codewords* must selectively hybridize in a manner that is compatible with the goals of the computation, and unwanted or non-selective hybridizations do not occur. In this section, we provide basic constraints for the design of appropriate sets of oligonucleotides called *DNA codes* or *DNA languages*.

4.1.1 Free Energy and Melting Temperature

Two physical constraints on DNA codes are addressed that refer to the stability of DNA strands in solution. A *free energy constraint* for a DNA code C is one in which any two codewords x and y in C must have comparable Gibbs free energy. That is, there is a constant $\delta > 0$ so that for any two codewords x and y in C,

$$|\Delta G^{\circ}(x) - \Delta G^{\circ}(y)| \leq \delta . \tag{4.1}$$

Z. Ignatova et al., *DNA Computing Models*,
DOI: 10.1007/978-0-387-73637-2_4, © Springer Science+Business Media, LLC 2008

A *melting temperature constraint* is one in which all codewords in the DNA code are forced to have similar melting temperature. This allows hybridization of multiple DNA strands to proceed simultaneously. Another *melting temperature constraint* is that all codewords in the DNA code must have similar GC-content and as a result similar thermodynamical characteristics. The GC-*content* of a DNA strand is defined as the number of positions in which the string has symbols C or G. The melting temperature constraint aims to design a DNA code so that a temperature can be found which is well above the melting temperature of all non-Watson-Crick pairs (x, y^{RC}), $x \neq y$, and well below the melting temperature of all Watson-Crick complementary pairs (x, x^{RC}). Then the formation of Watson-Crick pairs is significantly more energetically favorable than all possible non-Watson-Crick pairs.

4.1.2 Distance

The Hamming distance measure is particularly useful for oligonucleotides of length of 10 nt or less as the DNA sugar-phosphate backbone can be considered perfectly rigid.

In typical applications of DNA computing, the strings encoding information of interest are of the same length. Let Σ be an alphabet and let $n \geq 1$ be an integer. A *block code* of length n over Σ is a language over Σ consisting of strings with the same length n. In particular, a *DNA block code* is a block code over the DNA alphabet Δ.

Let C be a DNA block code, and let ϕ be a morphic or anti-morphic involution on Δ^*. The *ϕ-Hamming distance* between two codewords x and y in C is defined as the number $d_H^\phi(x, y)$ of positions at which the strings x and $\phi(y)$ differ. Notice that if a codeword x equals its image $\phi(x)$, then $d_H^\phi(x, \phi(x)) = 0$. The *minimum ϕ-Hamming distance* of C is the minimum non-zero ϕ-Hamming distance $d_H^\phi(C)$ between any two codewords in C. If ϕ is the identity mapping, then the ϕ-Hamming distance amounts to the ordinary *Hamming distance* denoted as d_H. If ϕ is the reverse complementarity, the ϕ-Hamming distance provides the *reverse-complement Hamming distance* termed d_H^{RC}. In this case, the identity $x^{RC} = \phi(x) = y$ represents the fact that the single-stranded molecules x and y could bind to each other.

Theorem 4.1. *Let ϕ be a morphic or anti-morphic involution on Σ^*. The involution ϕ is an isometry for d_H^ϕ on Σ^*. That is, for all strings x and y of the same length over Σ,*

$$d_H^\phi(x, y) = d_H^\phi(\phi(x), \phi(y)) . \qquad (4.2)$$

Moreover, for all strings x and y of the same length over Σ,

$$d_H(x, y) = d_H^\phi(x, \phi(y)) \quad and \quad d_H(x, \phi(y)) = d_H(\phi(x), y) . \qquad (4.3)$$

Finally, if C is ϕ-closed (i.e., with each codeword x the code C also contains the string $\phi(x)$), then $d_H^\phi(C) = d_H(C)$.

Proof. The first two assertions are obvious. Let C be ϕ-closed, and let x and y be codewords in C. Then $d_H(x,y) = d_H^\phi(x, \phi(y))$ and so $d_H(C) \geq d_H^\phi(C)$. Conversely, $d_H^\phi(x,y) = d_H(x, \phi(y))$ and thus $d_H^\phi(C) \geq d_H(C)$.   □

Example 4.2. Let ϕ be the Watson-Crick complementarity. The DNA words $x = $ ATGCTA and $y = $ AAGCTA have Hamming distance $d_H(x,y) = 1$ and ϕ-Hamming distance $d_H^\phi(x,y) = d_H(x, \phi(y)) = 3$, since $\phi(y) = $ TAGCTT.   ◇

4.1.3 Similarity

When DNA strands are considered as perfectly rigid, the Hamming distance is quite appropriate. However, for DNA strands of length of 10 nt or longer, it is more reasonable to consider the strands to be perfectly elastic like rubber-bands. For perfectly elastic strands it is possible for residues that are not necessarily at the same position in two strands to pair with each other.

Levenshtein Distance

Let Σ be an alphabet. The *Levenshtein distance*, also called *edit distance*, was first considered by V. Levenshtein in the mid-1960s. It is the distance between two strings x and y over Σ given by the minimum number $d_L(x,y)$ of operations needed to transform the string x into the string y, where the operations are the insertion and deletion of a single symbol. Insertion or deletion are referred to by the generic term *indel*.

Example 4.3. We have $d_L(\text{AGGT}, \text{GAT}) = 3$ as the following indels illustrate: AGGT (delete A), GGT (delete G), and GT (insert A), GAT.   ◇

Levenshtein also proposed a more complex edit distance measure using indels and the operation of substitution of single symbols. Then the edit distance can be considered as a generalization of the Hamming distance, which is only used for strings of the same length and only considers substitutions.

The edit distance between two strings is strongly related to their longest common subsequence. To see this, let $x = a_1 \ldots a_m$ and $y = b_1 \ldots b_n$ be strings over Σ. A string $z = c_1 \ldots c_l$ is called a *common subsequence* of x and y if two sequences of indices $1 \leq i_1 < \ldots < i_l \leq m$ and $1 \leq j_1 < \ldots < j_l \leq n$ exist so that $c_k = a_{i_k} = b_{j_k}$ for $1 \leq k \leq l$. Since the empty string is always a common subsequence, the set of common subsequences of any two strings is not empty. The *deletion similarity* between x and y is defined as the length $\ell(x,y)$ of the longest common subsequence of x and y.

Theorem 4.4. *The edit distance forms a metric on Σ^*, and for any two strings x and y over Σ,*

$$d_L(x, y) = |x| + |y| - 2\ell(x, y) . \tag{4.4}$$

Proof. Consider a graph whose vertices are the strings in Σ^*, with an edge joining the vertices corresponding to strings x and y, $x \neq y$, if and only if y can be obtained from x by an indel. Then the edit distance $d_L(x, y)$ is the length of the shortest path from x to y, and is thus indeed a metric. The second assertion is clear. $\square$

Consequently, the deletion similarity is related to the maximum number of Watson-Crick base pairs that may be formed between two anti-parallel strands. For two reverse complementary strands this number is simply their length.

Similarity Functions

The mathematical analysis of DNA strands can be based on similarity functions that measure thermodynamic similarity on single-stranded DNA. For this, let Σ be an alphabet. A *similarity function* on Σ^* is a mapping $\sigma : \Sigma^* \times \Sigma^* \to \mathbb{R}_0^+$ that satisfies the conditions

$$\sigma(x, x) \geq \sigma(x, y) = \sigma(y, x) \geq 0, \quad x, y \in \Sigma^* . \tag{4.5}$$

The Hamming distance gives rise to a similarity function $\sigma_\alpha : \Sigma^n \times \Sigma^n \to \mathbb{N}_0$ that assigns to each pair of strings x and y in Σ^n the number of positions $\sigma_\alpha(x, y)$ at which they coincide. This function is called *Hamming similarity* and fulfills the condition

$$\sigma_\alpha(x, y) = n - d_H(x, y), \quad x, y \in \Sigma^n . \tag{4.6}$$

The Levenshtein distance also provides a similarity function.

Theorem 4.5. *Let ϕ be a morphic or anti-morphic involution on Σ^*. A similarity function on Σ^* is given by*

$$\sigma_\lambda(x, y) = \ell(x, \phi(y)), \quad x, y \in \Sigma^* . \tag{4.7}$$

Example 4.6. Let ϕ be the reverse complementarity. Take the ϕ-closed DNA code $C = \{x, x^{RC}, y, y^{RC}\}$ with $x = $ AGAT and $y = $ ATAG, and thus $x^{RC} = $ ATCT and $y^{RC} = $ CTAT. The sequence AT is the longest common subsequence between any pair of distinct strings in C. Thus for any non-Watson-Crick pairing

$$\sigma_\lambda(x,x) = \sigma_\lambda(x^{RC}, x^{RC}) = 2,$$
$$\sigma_\lambda(y,y) = \sigma_\lambda(y^{RC}, y^{RC}) = 2,$$
$$\sigma_\lambda(x,y) = \sigma_\lambda(x^{RC}, y^{RC}) = 2,$$
$$\sigma_\lambda(y,x) = \sigma_\lambda(y^{RC}, x^{RC}) = 2,$$

while for the two Watson-Crick complementary pairings

$$\sigma_\lambda(x, x^{RC}) = \sigma_\lambda(x^{RC}, x) = 4,$$
$$\sigma_\lambda(y, y^{RC}) = \sigma_\lambda(y^{RC}, y) = 4 .$$

$\Diamond$

A similarity function related to the nearest-neighbor thermodynamics is based on the longest common block subsequence. For this, let $x = a_1 \ldots a_m$ and $y = b_1 \ldots b_n$ be strings over Σ. A common subsequence $z = c_1 \ldots c_l$ of x and y is called a *common block subsequence* if any two consecutive symbols in z are either consecutive in both x and y or are non-consecutive in both x and y. This means that the substring z can be defined by sequences of indices $1 \le i_1 < \ldots < i_l \le m$ and $1 \le j_1 < \ldots < j_l \le n$ so that $i_k + 1 = i_{k+1}$ is equivalent to $j_k + 1 = j_{k+1}$ for $1 \le k \le l - 1$. The *block similarity* between x and y is defined as the length $\sigma_\beta(x, y)$ of the longest common block subsequence of x and y. Clearly,

$$\sigma_\beta(x,y) \le \sigma_\lambda(x,y) \quad \text{and} \quad \sigma_\beta(x,x) = \sigma_\lambda(x,x) = |x| . \tag{4.8}$$

Example 4.7. The strings AGACT and AGT have AGT as the longest common subsequence, but AG as the longest common block subsequence. $\Diamond$

Theorem 4.8. *Let ϕ be a morphic or anti-morphic involution on Σ^*. The involution ϕ is an isometry for the similarities σ_β and σ_λ on Σ^*. That is, for all strings x and y over Σ,*

$$\sigma_\beta(x,y) = \sigma_\beta(\phi(x), \phi(y)) \quad \text{and} \quad \sigma_\lambda(x,y) = \sigma_\lambda(\phi(x), \phi(y)) . \tag{4.9}$$

A DNA block code C of length n is called a *reverse-complement (RC) closed DNA code* provided that each codeword occurs with its reverse complement and no codeword equals its reverse complement. That is, if $x \in C$ then $x^{RC} \in C$ and $x \ne x^{RC}$. Let σ be a similarity function on Δ^*. An RC-closed DNA code C is called a (n, D) *code based on the similarity* σ if for all codewords x, y in C with $x \ne y$,

$$\sigma(x,y) \le n - D - 1 . \tag{4.10}$$

In this case, we say that C is an RC-closed DNA code of length n, distance D, and deletion similarity $n - D - 1$.

Example 4.9. The DNA code C in Example 4.6 is RC-closed with length $n = 4$, distance $D = 1$, and deletion similarity $n - D - 1 = 2$.   ◇

4.2 DNA Languages

This section provides a formal language approach to constructing DNA codes that satisfy certain hybridization constraints.

4.2.1 Bond-Free Languages

A basic attempt to address the negative word design problem is to consider bond-free languages first studied by L. Kari and coworkers (2005).

Let Σ be an alphabet, and let $k \geq 0$ be an integer. Let ϕ denote a morphic or anti-morphic involution on Σ^*, and let L be a language over Σ. A *substring* of L is a string u over Σ that is a substring of a string w of L (i.e., $w = vuv'$ for some $v, v' \in \Sigma^*$). The language L is called (ϕ, k)-*bond-free* if, for any two substrings u and v of L with length k, we have that $u \neq \phi(v)$. Clearly, if L is (ϕ, k)-bond-free then L is (ϕ, k')-bond-free for each $k' \geq k$. Moreover, if L is (ϕ, k)-bond-free then each string in L is $\mathrm{hp}(\phi, k)$-free (Sect. 6.3.5).

Lemma 4.10. *Let S be the set of all subwords of length k in a language L over Σ. If $S \cap \phi(S) = \emptyset$, then the language L is (ϕ, k)-bond-free.*

Example 4.11. Let ϕ be the reverse complementarity. The DNA language $L = \{\mathtt{AC}, \mathtt{AA}\}$ is $(\phi, 1)$-bond-free, since $\phi(L) = \{\mathtt{GT}, \mathtt{TT}\}$.   ◇

In typical DNA applications, the strings encoding information of interest are usually obtained by concatenating fixed-length strings. For this, let K be a block code of length k over Σ. The set K^+ consists of all strings that are obtained by concatenating one or more strings from K. Each subset of K^+ is termed a *k-block code*. Let S be a set of strings of length k over Σ. The *k-substring closure* of S is the set $S^\otimes$ of all strings w of length at least k over Σ so that any substring of w of length k belongs to S.

Example 4.12. Consider the DNA block code $K = \{\mathtt{AA}, \mathtt{AC}, \mathtt{CA}, \mathtt{CC}\}$, and let ϕ be the reverse complementarity. Take the 2-block code $L = K^+$. Clearly, K is the set of all subwords of length 2 in L and thus L is the 2-substring closure of K. Moreover, $K \cap \phi(K) = \emptyset$ and hence Lemma 4.10 implies that the language L is $(\phi, 2)$-bond-free.

However, any extension of the language L by a string $w \in \Delta^2 \setminus K$ yields a language $L' = (K \cup \{w\})^+$, which is not $(\phi, 2)$-bond-free. For instance, if $w = \mathtt{AG}$, then $\mathtt{GC}$ is a subword in L' with the property $\mathtt{GC} = \phi(\mathtt{GC})$.

It is possible to construct a superset of L which is $(\phi, 2)$-bond-free. For this, add AG as a subword with the constraint that it cannot be followed by CA or CC. Moreover, add GA as a subword with the constraint that it cannot be preceded by AC, CC, or AG. The resulting language is $(\phi, 2)$-bond-free and can be accepted by a finite state machine. $\diamondsuit$

The situation described in the previous example will be generalized by the following

Theorem 4.13. *Let ϕ be the reverse complementarity, let $k \geq 1$ be an integer, and let K be the set of all strings of length k over $\{A, C\}$. The k-block code K^+ is the k-substring closure of K and is (ϕ, k)-bond-free.*

Bond-freedom can be generalized by making use of the Hamming distance. A language L over Σ is called $(\phi, H_{d,k})$-*bond-free* if for any two substrings u and v of length k in L, we have $d_H(u, \phi(v)) > d$. By definition, a language is $(\phi, H_{0,k})$-bond-free if and only if it is (ϕ, k)-bond-free.

Theorem 4.14. *Let j and q be positive integers, and let L be a subset of $\Sigma^{jq}\Sigma^*$. If L is $(\phi, H_{t,q})$-bond-free for some integer $t \geq 0$, then L is also $(\phi, H_{d,k})$-bond-free, where $d = j(t+1) - 1$ and $k = jq$.*

Proof. Let $u = u_1 \ldots u_j$ and $v = v_1 \ldots v_j$ be substrings of length k in L, where u_i and v_i are strings of length q, $1 \leq i \leq j$. By hypothesis, $d_H(u_i, \phi(v_i)) \geq t+1$ and thus $d_H(u, \phi(v)) = \sum_i d_H(u_i, \phi(v_i)) \geq j(t+1)$. $\square$

In particular, if $t = 0$ then every language that is $(\phi, H_{0,q})$-bond-free is also $(\phi, H_{d,(d+1)q})$-bond-free for any integer $d \geq 0$.

Theorem 4.15. *Let L be a language over Σ, let ϕ be a morphic or anti-morphic involution on Σ^*, and let d and k be integers with $k > d \geq 0$. If the language L is $(\phi, H_{d,k})$-bond-free, then L is also $(\phi, H_{0,k-d})$-bond-free.*

Proof. Let u and v be subwords of length $k - d$ in L. There are subwords of length k in L so that $u' = u_1 u u_2$ and $v' = v_1 v v_2$ for appropriate subwords u_1, u_2, v_1, and v_2. First, let ϕ be a morphic involution. By hypothesis, $d + 1 \leq d_H(u', \phi(v')) = d_H(u_1, \phi(v_1)) + d_H(u, \phi(v)) + d_H(u_2, \phi(v_2))$. But the strings $u_1 u_2$ and $v_1 v_2$ have length d and thus $d_H(u, \phi(v)) \geq 1$. The proof is similar if ϕ is an anti-morphic involution. $\square$

4.2.2 Hybridization Properties

Further algebraic properties of DNA languages are discussed that avoid certain undesirable hybridizations (Fig. 4.1) as introduced by L. Kari and coworkers (2006). Let Σ be an alphabet and let ϕ be a morphic or anti-morphic involution on Σ^*.

AGTA	AGTA	AGTA	AGTA
GCTCATTG	GTCATG	TCATG	GTCAT
(a)	(b)	(c)	(d)

Fig. 4.1 Different types of intermolecular hybridizations: (a) The codeword AGTA is the reverse complement of a subword given by the concatenation of two codewords, GCTC and ATTG; ϕ-comma-free codes avoid such hybridizations. (b) The codeword AGTA is the reverse complement of a subword of another codeword; ϕ-infix-free codes avoid such hybridizations. (c) The codeword AGTA is the reverse complement of a prefix of another codeword; ϕ–prefix-free codes avoid such hybridizations. (d) The codeword AGTA is the reverse complement of a suffix of another codeword; ϕ-suffix-free codes avoid such hybridizations.

- A language L over Σ is called ϕ-infix-free if $\Sigma^+\phi(L)\Sigma^* \cap L = \emptyset$ and $\Sigma^*\phi(L)\Sigma^+ \cap L = \emptyset$.
- A language L over Σ is termed ϕ-prefix-free if $\phi(L)\Sigma^+ \cap L = \emptyset$.
- A language L over Σ is called ϕ-suffix-free if $\Sigma^+\phi(L) \cap L = \emptyset$.
- A language L over Σ is termed ϕ-bifix-free if L is both ϕ-prefix-free and ϕ-suffix-free.
- A language L over Σ is called ϕ-comma-free if $\Sigma^+\phi(L)\Sigma^+ \cap L^2 = \emptyset$.

We say that a language L containing the empty word ϵ has one of the above properties if the language $L \setminus \{\epsilon\}$ has this property. A language L over Σ is called ϕ-strict if $L' \cap \phi(L') = \emptyset$, where $L' = L \setminus \{\epsilon\}$. If a language has one of the above properties and is strict, then the qualifier *strictly* is added.

Example 4.16. Let ϕ denote the reverse complementarity. The DNA language $L = \{\text{AGA}, \text{AC}\}$ with $\phi(L) = \{\text{TCT}, \text{GT}\}$ is ϕ-infix-free, while the DNA language $L' = \{\text{AGTC}, \text{AC}\}$ with $\phi(L') = \{\text{GTCT}, \text{GT}\}$ is not ϕ-infix-free, since AGTC $\in L$ belongs to $\Delta^+\phi(\text{AC})\Delta^+ = \Delta^+\text{GT}\Delta^+$.

The DNA language $L = \{\text{AG}, \text{AC}\}$ with $\phi(L) = \{\text{CT}, \text{GT}\}$ and $L^2 = \{\text{AGAG}, \text{AGAC}, \text{ACAG}, \text{ACAC}\}$ is ϕ-comma-free, while the DNA language $L = \{\text{AGC}, \text{TAC}\}$ with $\phi(L) = \{\text{GCT}, \text{GTA}\}$ is not ϕ-comma-free, since AGCTAC $\in L^2$ lies in $\Delta^+\phi(\text{AGC})\Delta^+ = \Delta^+\text{GCT}\Delta^+$. $\Diamond$

Lemma 4.17. *Let L be a language over Σ, and let ϕ be a morphic or anti-morphic involution on Σ^*.*

- *If L is ϕ-infix-free, then L is both ϕ-prefix-free and ϕ-suffix-free and thus ϕ-bifix-free.*
- *For a morphic involution ϕ, L is ϕ-prefix-free (suffix-free) if and only if $\phi(L)$ is ϕ-prefix-free (suffix-free).*
- *For an anti-morphic involution ϕ, L is ϕ-prefix-free (suffix-free) if and only if $\phi(L)$ is ϕ-suffix-free (prefix-free).*
- *L is ϕ-bifix-free if and only if $\phi(L)$ is ϕ-bifix-free.*
- *L is ϕ-comma-free if and only if $\phi(L)$ is ϕ-comma-free.*

Theorem 4.18. *For a morphic involution ϕ on Σ^*, the family of ϕ-prefix-free (ϕ-suffix-free) languages over Σ is closed under concatenation.*

Proof. Let L_1 and L_2 be ϕ-prefix-free languages. Suppose $L_1 L_2$ is not ϕ-prefix-free. Then there are strings $u_1 u_2, v_1 v_2 \in L_1 L_2$ so that $u_1 u_2 = \phi(v_1 v_2)w = \phi(v_1)\phi(v_2)w$ for some $w \in \Sigma^+$. But L_1 is ϕ-prefix-free, so $u_1 = \phi(v_1)$. It follows that $\phi(v_2)$ is a prefix of u_2, contradicting the assumption that L_2 is ϕ-prefix-free. □

Corollary 4.19. *Let ϕ be a morphic involution on Σ^*, and let $n \geq 1$ be an integer. If L is a ϕ-prefix-free (ϕ-suffix-free) language over Σ, then L^n is ϕ-prefix-free (ϕ-suffix-free).*

Theorem 4.20. *Let ϕ be a morphic or anti-morphic involution on Σ^*, and let $n \geq 1$ be an integer. If L is a ϕ-bifix-free language over Σ, then L^n is a ϕ-bifix-free language.*

Proof. If ϕ is a morphic involution, the result follows from Corollary 4.19. If ϕ is an anti-morphic involution, then ϕ-prefixes and ϕ-suffixes interchange. □

Theorem 4.21. *Let ϕ be a morphic or anti-morphic involution on Σ^*. If L is a strictly ϕ-infix-free language over Σ, then L^+ is both ϕ-prefix-free and ϕ-suffix-free.*

Proof. Suppose L^+ is not ϕ-prefix-free. Thus, $u_1 \ldots u_m = \phi(v_1 \ldots v_n)w$ for some $u_1, \ldots, u_m \in L$, $v_1, \ldots, v_n \in L$, and $w \in \Sigma^+$. For a morphic involution ϕ, we obtain $u_1 \ldots u_m = \phi(v_1) \ldots \phi(v_n)w$ and hence either $u_1 = \phi(v_1)$, or u_1 is prefix of $\phi(v_1)$, or $\phi(v_1)$ is prefix of u_1. All three conditions contradict the assumption that L is strictly ϕ-infix-free. For an anti-morphic involution ϕ, $u_1 \ldots u_m = \phi(v_n) \ldots \phi(v_1)w$ and the argument is similar. The assertion on ϕ-suffix-freedom can be analogously proved. □

Theorem 4.22. *Let ϕ be a morphic or anti-morphic involution on Σ^*. If L is a ϕ-comma-free language over Σ, then L is ϕ-infix-free.*

Proof. Let L be ϕ-comma-free over Σ. Suppose L is not ϕ-infix-free. We may assume without restriction that $u\phi(x)v = y$ for some $u \in \Sigma^*$, $v \in \Sigma^+$, and $x, y \in L$. Thus $y^2 = u\phi(x)vu\phi(x)v$ and hence L is not ϕ-comma-free. A contradiction. □

4.2.3 Small DNA Languages

Small DNA languages with predefined properties can be found by heuristic methods. Here a heuristics employed by F. Barany and coworkers (1999) is described which allows the construction of a set of oligonucleotides of length 24 nt. These oligonucleotides are termed *zip-codes*, as they are used in a DNA microarray that combines PCR and ligase detection reaction (LDR) with zip-code hybridization. Each zip-code is composed of six tetramers so that the full-length 24-mers have similar melting temperatures. First, the set of 256 ($= 4^4$) possible combinations in which the four nucleotides can be arranged as tetramers were reduced to a set of 36 tetramers. These tetramers were chosen so that they differ pairwise by at least two nucleotides. Moreover, tetramer complements and tetramers that are equal to their reverse complement, like TCGA, or repetitive, like CACA, were excluded. Furthermore, tetramers that are either all A and T bases or all G and C bases were deleted. The resulting 36 tetramers were the following:

> TTGA, TTAG, TCTG, TCCC, TCGT, TGTC, TGCG, TGAT, TACA,
> CTTG, CTCA, CTGT, CCTA, CCAT, CGTT, CGAA, CACG, CAGC,
> GTCT, GTGC, GCTT, GCAA, GGTA, GGAC, GATC, GACC, GAGT,
> ATCG, ATAC, ACCT, ACGG, AGTG, AGCC, AGGA, AATC, AAAG.

Six tetramers were selected from this set for use in designing the zip-codes. These tetramers differ from one another by at least two symbols:

> ACCT, ATCG, CAGC, GACC, GGTA, TGCG .

These six tetramers were combined so that each zip-code differs from all others by at least three alternating tetramers. This ensures that each zip-code differs from all other zip-codes by at least six symbols. The resulting zip-codes were the following:

$$5' - \text{TGCG} \,|\, \text{ACCT} \,|\, \text{CAGC} \,|\, \text{ATCG} \,|\, \text{ACCT} \,|\, \text{CAGC} - 3',$$
$$5' - \text{CAGC} \,|\, \text{ACCT} \,|\, \text{GACC} \,|\, \text{ATCG} \,|\, \text{ATCG} \,|\, \text{CAGC} - 3',$$
$$5' - \text{GACC} \,|\, \text{ACCT} \,|\, \text{TGCG} \,|\, \text{ATCG} \,|\, \text{GGTA} \,|\, \text{CAGC} - 3',$$
$$5' - \text{TGCG} \,|\, \text{GGTA} \,|\, \text{CAGC} \,|\, \text{ACCT} \,|\, \text{ACCT} \,|\, \text{TGCG} - 3',$$
$$5' - \text{CAGC} \,|\, \text{GGTA} \,|\, \text{GACC} \,|\, \text{ACCT} \,|\, \text{ATCG} \,|\, \text{TGCG} - 3',$$
$$5' - \text{GACC} \,|\, \text{GGTA} \,|\, \text{TGCG} \,|\, \text{ACCT} \,|\, \text{GGTA} \,|\, \text{TGCG} - 3',$$
$$5' - \text{TGCG} \,|\, \text{ATCG} \,|\, \text{CAGC} \,|\, \text{GGTA} \,|\, \text{ACCT} \,|\, \text{GACC} - 3',$$
$$5' - \text{CAGC} \,|\, \text{ATCG} \,|\, \text{GACC} \,|\, \text{GGTA} \,|\, \text{ATCG} \,|\, \text{GACC} - 3',$$
$$5' - \text{GACC} \,|\, \text{ATCG} \,|\, \text{TGCG} \,|\, \text{GGTA} \,|\, \text{GGTA} \,|\, \text{GACC} - 3' .$$

4.3 DNA Code Constructions and Bounds

Several constructions and bounds on DNA block codes are described which are based on Hamming distance and deletion similarity.

4.3.1 Reverse and Reverse-Complement Codes

Algebraic coding provides upper and lower bounds on the error correction capabilities of block codes. For this, let Σ_q be an alphabet with q elements. A block code over Σ_q of length n with M codewords and minimum Hamming distance d is called (n, M, d, q) *code*, or simply (n, M, d) *code* if the alphabet is known. Moreover, the maximum size of a block code over Σ_q of length n with minimum Hamming distance d is denoted by $A_q(n, d)$.

In the following, bounds on the size of DNA block codes are derived, based on the work of R. Corn and coworkers (2001). For this, a (n, M, d) code is called a *reverse code* if for any pair of codewords x and y, $d_H(x, y) \geq d$, if $x \neq y$, and $d_H(x, y^R) \geq d$. Let $A_q^R(n, d)$ denote the maximum size of a reverse code over Σ_q of length n and minimum Hamming distance d. The reverse constraint limits hybridization between a codeword and the reverse of a codeword.

A (n, M, d) code is called a *reverse-complement code* or *RC code* if for all pairs of codewords x and y, $d_H(x, y) \geq d$, if $x \neq y$, and $d_H(x^C, y^R) \geq d$. By Theorem 4.1, the latter inequality is equivalent to $d_H(x, y^{RC}) \geq d$. Let $A_q^{RC}(n, d)$ denote the maximum size of an RC code over Σ_q of length n and minimum Hamming distance d. The reverse-complement constraint limits hybridization between a codeword and the reverse complement of a codeword. There is a close relationship between the maximum sizes of reverse and reverse-complement codes.

Theorem 4.23. *Let $n \geq 1$ be an integer. If n is even, then*

$$A_4^{RC}(n, d) = A_4^R(n, d) , \tag{4.11}$$

and if n is odd then

$$A_4^{RC}(n, d) \leq A_4^R(n + 1, d + 1) . \tag{4.12}$$

Proof. Let n be even. Let C be a (n, M, d) reverse code. Write each codeword of C in the form $c = ab$ so that a and b have equal length. Claim that $C' = \{ab^C \mid ab \in C\}$ is an RC code. Indeed, in view of Theorem 4.1,

$$d_H(ab^C, uv^C) = d_H(a, u) + d_H(b^C, v^C) = d_H(a, u) + d_H(b, v)$$
$$= d_H(ab, uv) .$$

Moreover by Theorem 4.1,

$$d_H(ab^C, (uv^C)^{RC}) = d_H(ab^C, v^R u^{RC}) = d_H(a, v^R) + d_H(b^C, u^{RC})$$
$$= d_H(a, v^R) + d_H(b, u^R) = d_H(ab, (uv)^R) .$$

It follows that $A_4^R(n, d) \leq A_4^{RC}(n, d)$. The reverse inequality can be similarly proved.

Let n be odd. Let C be a $(n+1, M, d+1)$ reverse code. Puncture the code C by deleting the center position. Assume that the center position is informative in the sense that at least two codewords differ at the center position. The resulting code is a (n, M', d) reverse code with $M' \leq M$. So the result follows from the previous assertion for n even. $\square$

Hence, in order to derive bounds on the maximum size of RC codes, it is sufficient to consider reverse codes.

Example 4.24. The DNA code C in Example 4.6 is a $(4, 4, 2)$ reverse code. The proof of Theorem 4.23 shows that the code $C' = \{\texttt{AGTA}, \texttt{ATTC}, \texttt{ATGA}, \texttt{CTTA}\}$ is a $(4, 4, 2)$ RC code. $\diamond$

Theorem 4.25 (Johnson-Type Bound). *Let $n \geq 1$ be an odd integer. For each integer d with $0 \leq d \leq n$,*

$$A_4^R(n, d) \leq \lfloor \frac{1}{4} A_4^R(n - 1, d) \rfloor .\tag{4.13}$$

Proof. Let C be a (n, M, d) reverse code. By the pigeonhole principle, there are at least $\lceil M/4 \rceil$ codewords which contain in the center position the same symbol. By keeping just these codewords and deleting this position, the resulting reverse code has length $n - 1$ and minimum Hamming distance at least d. $\square$

Theorem 4.26 (Halving Bound). *Let $n \geq 1$ be an integer. For each integer d with $0 < d \leq n$,*

$$A_4^R(n, d) \leq \frac{1}{2} A_4(n, d) .\tag{4.14}$$

Proof. Let C be a (n, M, d) reverse code. By hypothesis $d > 0$ and so the sets C and $C^R = \{c^R \mid c \in C\}$ are disjoint. Thus $C' = C \cup C^R$ is a $(n, 2M, d)$ code. Indeed, Theorem 4.1 shows that for each pair of codewords $x, y \in C$, $d_H(x^R, y^R) = d_H(x, y)$ and $d_H(x^R, (y^R)^R) = d_H(x, y^R)$. $\square$

Example 4.27. The DNA code C in Example 4.6 is a $(4, 4, 2)$ reverse code. The proof of the halving bound shows that the following DNA code is a $(4, 8, 2)$ code:

$$C' = \{\texttt{AGAT}, \texttt{ATAG}, \texttt{ATCT}, \texttt{CTAT}, \texttt{TAGA}, \texttt{GATA}, \texttt{TCTA}, \texttt{TATC}\} .$$

$\diamond$

Theorem 4.28 (Product Bound). *Let $n \geq 1$ be an integer. For all integers d and w with $0 \leq d \leq n$ and $0 \leq w \leq n$,*

$$A_4^R(n, d) \geq A_2^R(n, d) \cdot A_2(n, d) .\tag{4.15}$$

Proof. Let C_1 be a binary (n, M_1, d) reverse code, and let C_2 be a binary (n, M_2, d) code. Consider the DNA code $C = C_1 \odot C_2 = \{x \odot y \mid x \in C_1, y \in C_2\}$ of length n (Sect. 4.3.2). Clearly, the code C has minimum Hamming distance d. Moreover, $d_H(a \odot b, (u \odot v)^R) = d_H(a \odot b, u^R \odot v^R) \geq d_H(a, u^R)$, as required. $\square$

Example 4.29. The code $C_1 = \{1010, 1100\}$ is a binary $(4, 2, 2)$ reverse code, and the code $C_2 = \{0000, 1100\}$ is a binary $(4, 2, 2)$ code. The proof of the product bound exhibits that the code $C = C_1 \odot C_2$ is a DNA $(4,2,2)$ reverse code with the codewords

$$\text{CACA} = 1010 \odot 0000,$$
$$\text{CCAA} = 1100 \odot 0000,$$
$$\text{GTCA} = 1010 \odot 1100,$$
$$\text{GGAA} = 1100 \odot 1100.$$

4.3.2 Constant GC-Content Codes

Bounds for DNA codes with constant GC-content are described, based on the work of O. King (2003). To this end, define a mapping $\odot$ from pairs of binary strings of length n to DNA strings of length n, given by $x \odot y = z$, where $z_i = \text{A}$ if $x_i = 0$ and $y_i = 0$, $z_i = \text{C}$ if $x_i = 1$ and $y_i = 0$, $z_i = \text{G}$ if $x_i = 1$ and $y_i = 1$, and $z_i = \text{T}$ if $x_i = 0$ and $y_i = 1$. For instance, we have $01100 \odot 01010 = \text{AGCTA}$. This mapping is bijective, and a DNA block code C factors into an *even component* $E(C) = \{x \mid x \odot y \in C\}$ and an *odd component* $O(C) = \{y \mid x \odot y \in C\}$. Conversely, if E and O are binary codes of length n, then the set $C = E \odot O = \{x \odot y \mid x \in E, y \in O\}$ is an DNA code of length n so that E is the even component and O is the odd component of the code C.

Define the *Hamming weight* of a codeword as the number of non-zero components. A code is termed *constant-weight* if all codewords have the same Hamming weight. Notice that the GC-content of a codeword $z = x \odot y$ equals the Hamming weight of its even component x. Therefore, we have the following:

Lemma 4.30. *A DNA block code is a constant GC-content code if and only if its even component forms a constant-weight code.*

Example 4.31. The DNA code constructed in Example 4.29 is a $(4,4,2)$ reverse code with GC-content $w = 2$. ◇

Example 4.32. Let X be a set of n elements called *points*. A *Steiner system* is a set of k-subsets, termed *blocks*, of X with the property that any t-subset of X is contained in exactly one block. This Steiner system is denoted by $S(t, k, n)$. Each block in a Steiner system can be viewed as an incidence vector of length n. In this way, the Steiner system forms a binary block code of length n given by the blocks as codewords. The total number of blocks in $S(t, k, n)$ providing the number of codewords is given by

$$M = \binom{n}{t} / \binom{k}{t}.$$

Every codeword has Hamming weight k and so the code is a constant-weight code.

For instance, the *projective plane of order 2* is formed by seven points and seven lines. The lines are 124, 235, 346, 457, 156, 267, and 137. These blocks form a Steiner system $S(2, 3, 7)$ and so provide a binary constant-weight code of length $n = 7$ consisting of $M = 7$ codewords. $\diamond$

Define the quantities $A_4^{GC}(n, d, w)$, $A_4^{GC,R}(n, d, w)$, and $A_4^{GC,RC}(n, d, w)$ in the same way as the respective quantities $A_4(n, d)$, $A_4^R(n, d)$, and $A_4^{RC}(n, d)$, but with the additional requirement that each codeword has GC-content $w \geq 0$.

Theorem 4.33. *Let $n \geq 1$ be an integer, and let d and w be integers with $0 \leq d \leq n$ and $0 \leq w \leq n$. If n is even then*

$$A_4^{GC,RC}(n, d, w) = A_4^{GC,R}(n, d, w), \tag{4.16}$$

and if n is odd then

$$A_4^{GC,RC}(n, d, w) \leq A_4^{GC,R}(n + 1, d + 1, w). \tag{4.17}$$

The proof is similar to that for DNA codes with unrestricted GC-content.

Theorem 4.34 (Johnson-Type Bound). *Let $n \geq 1$ be an integer. For all integers d and w with $0 \leq d \leq n$ and $0 < w < n$,*

$$A_4^{GC}(n, d, w) \leq \lfloor \frac{2n}{w} A_4^{GC}(n - 1, d, w - 1) \rfloor, \tag{4.18}$$

$$A_4^{GC}(n, d, w) \leq \lfloor \frac{2n}{n - w} A_4^{GC}(n - 1, d, w) \rfloor. \tag{4.19}$$

Proof. Let C be a (n, M, d) code with constant GC-content w. By the pigeon-hole principle, there is a position j in which at least $\lceil wM/2n \rceil$ codewords have nucleotide G, or there is a position j in which at least $\lceil wM/2n \rceil$ codewords have nucleotide C. Otherwise, the average GC-content would be less than w. By keeping just these codewords and deleting the jth position, the resulting code has length $n - 1$, GC-content $w - 1$, and minimum Hamming distance of at least d. The second assertion is proved similarly, using some position with at least $\lceil (n - w)M/2n \rceil$ A's or $\lceil (n - w)M/2n \rceil$ T's. $\square$

Theorem 4.35 (Halving Bound). *Let $n \geq 1$ be an integer. For all integers d and w with $0 < d \leq n$ and $0 \leq w \leq n$,*

$$A_4^{GC,R}(n, d, w) \leq \frac{1}{2} A_4^{GC}(n, d, w). \tag{4.20}$$

The proof is analogous to that for reverse codes with unconstrained GC-content.

Theorem 4.36 (Gilbert-Type Bound). *Let $n \geq 1$ be an integer. For all integers d and w with $0 \leq d \leq n$ and $0 \leq w \leq n$,*

$$A_4^{GC}(n, d, w) \geq \frac{\binom{n}{w} 2^w 2^{n-w}}{\sum_{k=0}^{d-1} \sum_{i=0}^{\min\{\lfloor k/2 \rfloor, w, n-w\}} \binom{w}{i} \binom{n-w}{i} \binom{n-2i}{k-2i} 2^{2i}} . \qquad (4.21)$$

Proof. The numerator provides the total number of DNA strings of length n with GC-content w. The denominator gives the number of these strings that have distance at most $d - 1$ from any fixed codeword x. In particular, $\binom{w}{i} \binom{n-w}{i} \binom{n-2i}{k-2i} 2^{2i}$ is the number of strings y with GC-content w, $d_H(x, y) = k$, and for which there are exactly $w - i$ positions in x and y that contain both G or C. □

Define the quantities $A_2(n, d, w)$ and $A_2^R(n, d, w)$ in the same way as the respective quantities $A_2(n, d)$ and $A_2^R(n, d)$, but with the additional requirement that each codeword has Hamming weight $w \geq 0$.

Theorem 4.37 (Product Bounds). *Let $n \geq 1$ be an integer. For all integers d and w with $0 \leq d \leq n$ and $0 \leq w \leq n$,*

$$A_4^{GC}(n, d, w) \geq A_2(n, d, w) \cdot A_2(n, d), \qquad (4.22)$$

$$A_4^{GC,R}(n, d, w) \geq A_2^R(n, d, w) \cdot A_2(n, d), \qquad (4.23)$$

$$A_4^{GC,R}(n, d, w) \geq A_2(n, d, w) \cdot A_2^R(n, d) . \qquad (4.24)$$

Proof. The second assertion will be proved, the other two assertions can be similarly shown. Let C_1 be a binary (n, M_1, d) reverse code with constant Hamming weight w, and let C_2 be a binary (n, M_2, d) code. Consider the DNA code $C = \{x \odot y \mid x \in C_1, y \in C_2\}$ of length n. Clearly, C has GC-content w and minimum Hamming distance d. Moreover, $d_H(a \odot b, (u \odot v)^R) = d_H(a \odot b, u^R \odot v^R) \geq d_H(a, u^R)$, as required. □

4.3.3 Similarity-Based Codes

This section provides bounds on the size of RC-closed DNA codes based on the recent work of A. D'yachkov and coworkers. For this, let $N_4(n, D)$ denote the maximum size of an RC-closed DNA (n, D) code related to a given similarity σ. Notice that $N_4^\beta(n, D)$ and $N_4^\lambda(n, D)$ are analogously defined, where the upper indices indicate that the corresponding codes are based on the similarity functions σ_β and σ_λ, respectively. As common block subsequences are common subsequences per se, it follows that $N_4^\lambda(n, D) \leq N_4^\beta(n, D)$.

Let us start with an upper bound on the code size by using a well-known argument from algebraic coding.

Theorem 4.38 (Hamming Bound). *Let $n \geq 1$ be an integer. We have*

$$N_4^\beta(n,1) \leq (4^{n-1} + 4)/2 \,, \tag{4.25}$$

and for any integer D, $2 \leq D \leq n$,

$$N_4^\beta(n,D) \leq \frac{4^n}{\sum_{i=0}^{\lfloor D/2 \rfloor} \binom{n}{i} 3^i} \,. \tag{4.26}$$

Proof. Consider an RC-closed DNA $(n,1)$ code C with M codewords. Each codeword x has one or two different tail subsequences of length $n-1$, which are obtained by deleting the first or the last symbol from x. Let M_1 denote the number of codewords in C which have one tail subsequence, and let M_2 denote the number of codewords in C with two tail subsequences. Then $M = M_1 + M_2$.

A codeword in which the tail subsequences coincide is composed of a single symbol. Thus, $M_1 \leq 4$. Moreover, for distinct codewords x and y in C, we have by definition $\sigma_\beta(x,y) \leq n-2$ and thus the resulting $M_1 + 2M_2$ strings of length $n-1$ are all distinct. Hence, $M_1 + 2M_2 \leq 4^{n-1}$ and the result follows.

Let $D \geq 2$. Take a DNA block code C of length n and Hamming distance $D+1$. Put $r = \lfloor D/2 \rfloor$ and provide each codeword $x \in C$ with a sphere $B_4(x) = \{y \mid d_H(x,y) \leq r\}$ of radius r centered at the codeword. The denominator in Eq. (4.26) provides the number of words in the sphere. For distinct codewords x and y in C, we have by hypothesis $d_H(x,y) \geq D+1$ and thus the spheres around distinct codewords are disjoint. Hence, the assertion follows.   □

Example 4.39. Consider the set of $4^3 = 64$ DNA codewords $x = a_1 \ldots a_4$ of length $n = 4$ satisfying the parity check condition $a_1 + \ldots + a_4 \equiv 0 \mod 4$ (Table 4.1).

This code contains an RC-closed $(4,1)$ code of 34 elements with block distance $D = 1$ and block deletion similarity 2 (by deleting the pairs marked by the symbol '−'). Hence, the upper bound $N_4^\beta(4,1) = 34$ in Theorem 4.38 is tight.

Moreover, the latter code contains an RC-closed $(4,1)$ code of 20 elements with distance $D = 1$ and deletion similarity 2 (by taking the pairs marked by the symbol '+'). Therefore, $N_4^\lambda(4,1) \geq 20$.   ◇

Finally, lower bounds on the code size are provided by using a random coding method. To this end, let $n \geq 1$ and s be integers with $0 \leq s \leq n$, and let σ be a similarity function on Δ^n. Define

$$P(n,s) = \{(x,y) \in \Delta^n \times \Delta^n \mid \sigma(x,y) = s\} \tag{4.27}$$

Table 4.1 A DNA block code of length 4 satisfying the parity check condition. The codewords are divided into pairs of mutually reverse complementary codewords. In each row, each pair is obtained from the first pair by a cyclic shift.

0000 3333+			
0013 0233	3001 2330-	1300 3302+	0130 3023-
0022 1133+	2002 1331-	2200 3311+	0220 3113-
0031 2033	1003 0332-	3100 3320	0310 3203-
0103 0323+	3010 3230-	0301 2303+	1030 3032-
0112 1223	2011 2231-	1201 2312+	1120 3122-
0121 2123	1012 1232-	2101 2321	1210 3212-
0202 1313	2020 3131-		
0211 2213+	1021 2132-	1120 1322+	2110 3221-
1111 2222+			

and

$$P^{RC}(n, s) = \{x \in \Delta^n \mid \sigma(x, x^{RC}) = s\} . \tag{4.28}$$

Consider random strings $x, y \in \Delta^n$ with independently identically distributed symbols having the uniform distribution on Δ. The corresponding random variables $\sigma(x, y)$ and $\sigma(x, x^{RC})$ have the respective probability distributions

$$\Pr\{\sigma(x, y) = s\} = \frac{|P(n, s)|}{4^{2n}} \tag{4.29}$$

and

$$\Pr\{\sigma(x, x^{RC}) = s\} = \frac{|P^{RC}(n, s)|}{4^n} . \tag{4.30}$$

Take the probabilities

$$P_1(n, D) = \Pr\{\sigma(x, x^{RC}) \geq n - D\} = \frac{1}{4^n} \sum_{t=0}^{D} |P^{RC}(n, n - t)| \tag{4.31}$$

and

$$P_2(n, D) = \Pr\{\sigma(x, y) \geq n - D\} = \frac{1}{4^{2n}} \sum_{t=0}^{D} |P(n, n - t)| . \tag{4.32}$$

The following result holds for both deletion similarities.

Theorem 4.40 (Random Coding Bound). *Let $n \geq 1$ be an integer. For any integer D, $1 \leq D \leq n - 1$,*

$$N_4(n, D) \geq \left\lfloor \frac{\frac{1}{2} - P_1(n, D)}{2P_2(n, D)} \right\rfloor + 1 . \tag{4.33}$$

Proof. Consider an RC-closed DNA code C of length n with codewords $x_1, \ldots, x_{2N}$ so that $x_i^{RC} = x_{N+i}$, $1 \le i \le N$. A codeword x_i is called *D-bad* in C if there is an index j with $j \ne i$ so that $\sigma(x_i, x_j) \ge n - D$. A pair of reverse complementary codewords (x_i, x_{N+i}) is *D-bad* in C provided that at least one of them is *D-bad*. This event is denoted by $E(i, D)$. If all pairs of codewords in C are *D-good* (i.e., not *D*-bad), then C is an RC-closed DNA (n, D) code.

Take an ensemble of these codes C so that all $n \times N$ components of the codewords $x_1, \ldots, x_N$ are independently identically distributed random variables with the uniform distribution on Δ. The distribution of the random variable $\sigma(x_i, x_j)$, $i \ne j$, has the form (4.29) or (4.30) depending on whether or not the equality $|i - j| = N$ is satisfied. By the additivity of independent events, we obtain the inequality

$$\Pr\{E(i, D)\} \le (2N - 2)P_2(n, D) + P_1(n, D) . \tag{4.34}$$

Put

$$N = \left\lfloor \frac{\frac{1}{2} - P_1(n, D)}{2P_2(n, D)} \right\rfloor + 1 . \tag{4.35}$$

Then Eq. (4.34) yields that the probability of the event $E(i, D)$, $1 \le i \le N$, does not exceed $1/2$. Thus, the average number of *D*-bad pairs in a code C from the ensemble does not exceed $\lfloor N/2 \rfloor$. Hence, there exists a code C with at least $\lfloor N/2 \rfloor$ *D*-good pairs. These *D*-good pairs form an RC-closed DNA (n, D) code of size at least $2\lfloor N/2 \rfloor \ge N - 2$. $\square$

Theorem 4.41. *If $D \ge 1$ is a fixed integer and $n \to \infty$, then*

$$N_4^\lambda(n, D) \ge (D!)^2 \left(\frac{4}{3^2} \right)^D \frac{4^n}{n^{2D}} (1 + o(1)) . \tag{4.36}$$

Proof. Let z be a string of length s over Δ, $0 \le s \le n$. Let $S_4(z, n)$ denote the set of all superstrings x of length n over Δ that have z as subsequence. The size of $S_4(z, n)$ is independent of z and is given by

$$|S_4(n, s)| = \sum_{k=0}^{n-s} \binom{n}{k} 3^k . \tag{4.37}$$

It follows that

$$|P^\lambda(n, s)| \le 4^s |S_4(n, s)|^2 \tag{4.38}$$

and if $s \ge 1$ is even,

$$|P^{RC,\lambda}(n, s)| \le 4^{s/2} |S_4(n, s)| . \tag{4.39}$$

Moreover, the set $P^{RC,\lambda}(n,s)$ is empty if $s \geq 1$ is odd. The bounds (4.38) and (4.39) can be used to establish the probabilities (4.31) and (4.32), respectively. The result then follows from the random coding bound.   □

Theorem 4.42. *If $D \geq 1$ is a fixed integer and $n \to \infty$, then*

$$N_4^\beta(n,D) \geq \frac{1}{4}\frac{D!}{4^D}\frac{4^n}{n^D}(1+o(1)).\tag{4.40}$$

Proof. Claim that for any integer s with $1 \leq s \leq n$, we have

$$|P^\beta(n,s)| \leq 4^s \sum_{k=1}^{\min\{s,n-s+1\}} \binom{s-1}{k-1}\left[4^{n-s}\binom{n-s+1}{k}\right]^2.\tag{4.41}$$

Indeed, for a fixed string z of length s over Δ, consider the block k-partition decomposing the symbols of z into k blocks, that is, $Z = (B_1,\ldots,B_k)$ with $|B_i| \geq 1$ and $\sum_i |B_i| = s$. Assume that z is a subsequence of a string x of length n over Δ. In this case, the partition B must be β-*matched* with x, that is, there exists an embedding of z into x so that each block B_i has no blanks in x and any two neighboring blocks are separated in x.

The number of sequences z of length s over Δ is 4^s. Consider all possible partitions Z with the number of blocks bounded by $k \leq \min\{s, n-s+1\}$. For a fixed integer k, the number of such partitions equals $\binom{s-1}{k-1}$. For each partition, we consider two sequences of length $n-s$ that complete z to strings x and y of length n. Each of these strings is divided into $k+1$ blocks that will be inserted between the corresponding blocks of the partition Z. The first and the last block may be empty, while all other blocks are non-empty. The number of such partitionings equals $\binom{n-s+1}{k}$. This proves the claim.

Clearly, the set $P^{RC,\beta}(n,s)$ is empty if $s \geq 1$ is odd. Moreover, if s is an even integer with $2 \leq s \leq n$,

$$|P^{RC,\beta}(n,s)| \leq 4^{s/2} \sum_{k=1}^{\min\{s,n-s+1\}} \binom{s/2-1}{\lceil k/2\rceil - 1}\left[4^{n-s}\binom{n-s+1}{k}\right].\tag{4.42}$$

This inequality can be similarly proved as the above assertion. The Eqs. (4.41) and (4.42) can be used to establish the probabilities (4.31) and (4.32), respectively. The result then follows from the random coding bound.   □

4.4 In Vitro Random Selection

In vitro evolution makes it possible to generate molecules with desired properties. In particular, libraries of non-crosshybridizing oligonucleotides can be evolved in a test tube.

4.4.1 General Selection Model

The design of biomolecules with defined structure and function is an unattained goal. Today, protein folding and catalysis are incompletely understood. Synthetic enzyme mimics that are rationally designed are still inefficient catalysts in comparison with natural enzymes. Instead of rational design, more and more evolutionary approaches are used to create biomolecules with the desired function. These methods require the preparation of a large library of DNA, RNA, or proteins from which those with desirable properties are selected and amplified. In such experiments, the molecules compete with each other and only those that succeed will be amplified. The amplified molecules are usually subject to a few rounds of selection and selective amplification so that in the end, the molecules with the best properties will be obtained. In every round, the selection pressure can be enhanced. Between two rounds, the pool of molecules may be mutated in order to increase the diversity of the present molecules. In this way, each round consists of mutation, selection, and amplification of molecules.

The first step involves the preparation of a library of compounds called the *seed set*. To this end, a DNA synthesizer will be programmed so that the single DNA strands have constant end sequences needed for primer recognition, but a randomized sequence in the middle. The mixture obtained consists of about 10^{15} to 10^{16} individual molecules.

This mixture can be put onto an affinity column, which is a column filled with a material (e.g., polymer or hydrophobic silica gel) onto which a guest molecule is specifically bound (Fig. 4.2). Those individual molecules that have some affinity to the column, and most likely to the compound bound to the surface, will be retarded. These molecules are eluted by increasing the salt concentration of the eluent. The corresponding fraction is collected and the contained molecules are amplified by PCR. In this way, a pool of compounds is obtained in which binding DNA or RNA molecules are enriched. This pool is then again applied to the affinity column and the whole process of selection and amplification is repeated. The selection pressure can be changed by eluting with a more shallow salt gradient and by collecting only compounds that come off the column at higher salt conditions. This will elute only the most strongly bound molecules. A special trick to find highly selective compounds is to randomize each selected pool again. This is achieved by modifying the PCR reaction using error-prone DNA polymerase to copy a DNA strand. The error rate can be tuned to be between one error per 1 to every 20 base pairs.

4.4.2 Selective Word Design

An in vitro evolutionary method that allows the design of DNA words for DNA computation was proposed by R. Deaton and coworkers (1996). To this

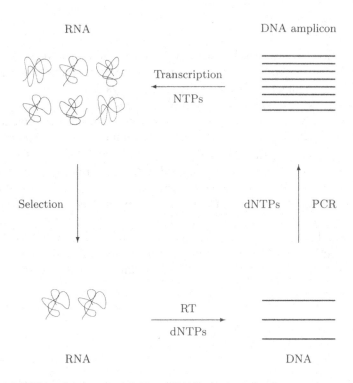

Fig. 4.2 RNA selection model: First, RNA is selected and copied into DNA using reverse transcriptase (RT). The obtained DNA is amplified and transcribed again into RNA before the next round of selection. All remaining DNA after each round of selection can be digested with DNAse. The finally selected RNA will be reverse-transcribed into DNA and the DNA will be sequenced.

end, the seed set consists of oligonucleotides of the same length, which have primer sequences at both ends and a randomized segment somewhere in the middle. Each round consists of heating the mixture to a high temperature and then rapidly cooling to room temperature or below. This *quenching step* during annealing should increase the likelihood that *highly mismatched molecules* are formed that may have complementary primer strands attached, but no complementary strand to the randomized segment in the middle. The formation of highly mismatched molecules was confirmed by experimental design. Then PCR is preformed at low temperature. This will selectively enhance highly mismatched molecules, while molecules closer to being Watson-Crick complements are not enhanced. In this way, a huge library of very mismatched DNA words can be selectively generated. This method has the advantage that the library is derived under conditions that are somewhat similar to that used for DNA computing.

Concluding Remarks

DNA codes that satisfy certain combinatorial constraints are very useful in DNA computing to guarantee the robustness of DNA computations. Appropriate DNA codes may be constructed in several ways. First, DNA codes may be designed by using a language-theoretic approach. This approach is largely definition-oriented and currently lacks constructive methods to derive larger DNA codes that simultaneously satisfy several combinatorial constraints. Second, methods from traditional coding theory can be adopted to design DNA codes. For this, quaternary codes with additional algebraic properties (e.g., linearity, cyclicality) may be studied by using the existing machinery of algebraic coding. This branch seems to be the most promising direction of DNA coding. Third, DNA codes may be developed by making use of similarity-based coding. This field, initiated by the pioneering work of V. Levenshtein, regained attention through the recent work of A. D'yachkov and coworkers. However, similarity-based coding is still in its infancy as an algebraic description is lacking.

References

1. Ackermann J, Gast FU (2003) Word design for biomolecular information processing. Zeitschr Naturforsch 58:157–161
2. Bours PAH (1995) On the construction of perfect deletion-correcting codes using design theory. Designs, Codes, Cryptography 6:5–20
3. Brouwer A (1998) In: Pless V, Huffman W (eds.) Handbook of coding theory. Elsevier, Amsterdam
4. Day WHE (1984) Properties of Levenshtein metrics on sequences. Bull Math Biology 46:327–332
5. Domaratzki M (2006) Bond-free DNA language classes. Nat Comput 6:371–402
6. Deaton R, Garzon R, Murphy J, Rose D, Franceshetti, Stevens Jr S (1996) Genetic search of reliable encodings for DNA based computation. In: Koza J, Goldberg D, Fogel R, Riolo R (eds.) First Conf Genetic Programming. Stanford, CA
7. D'yachkov A, Erdös P, Macula A, Rykov V, Vilenkin P, White S (2003) Exordium for DNA codes. J Combinat Opt 7:369–379
8. D'yachkov A, Macula A, Pogozelski W, Renz T, Rykov V, Torney D (2005) A weighted insertion–deletion stacked pair thermodynamic metric for DNA codes. LNCS 3384:90–103
9. D'yachkov A, Vilenkin P, Ismagilo K, Sarbaev RS, Macula A, Torney D, White S (2005) On DNA codes. Probl Inform Trans 41:349–367
10. Frutos A, Liu Q, Thiel A, Sanner AM, Condon A, Smith L, Corn R (1997) Demonstration of a word design strategy for DNA computing on surfaces. Nuc Acids Res 25:4748–4757
11. Gaborit P, King O (2005) Linear constructions for DNA codes. Theoret Comp Sci 334:99–113
12. Garzon M, Phan Y, Roy S, Neel A (2006) In search of optimal codes for DNA computing. LNCS 4287:143–156

13. Gerry NP, Witowski NE, Day J, Hammer RP, Barany G, Barany F (1999) Universal DNA microarray method for multiplex detection of low abundance point mutations. J Mol Biol 292:251–262

14. Kari L, Konstantinidis S, Sosik P (2005) Bond-free languages: formalizations, maximality and construction methods. LNCS 3384:169–181

15. Kari L, Konstantinidis S, Sosik P (2005) Preventing undesirable bonds between DNA codewords. LNCS 3384:182–191

16. Kari L, Konstantinidis S, Losseva E, Wozniak G (2003) Sticky-free and overhang-free DNA languages. Acta Inform 40:119–157

17. Kari L, Mahalingam K (2006) DNA codes and their properties. LNCS 4287: 127–142

18. Kao MY, Sanghi M, Schweller R (2005) Randomized fast design of short DNA words. LNCS 3580:1275–1286

19. King O (2003) Bounds for DNA codes with constant GC-content. Elect J Combinat 10:1–13

20. Levenshtein V (1966) Binary codes capable of correcting deletions, insertions, and reversals. Soviet Physics Dokl 10:707–710

21. Levenshtein V (2001) Efficient reconstruction of sequences. IEEE Trans Inform Theory 47:2–22

22. Levenshtein V (2002) Bounds for deletion/insertion correcting codes. Proc ISIT 2002, Lausanne, Switzerland

23. Litsyn S (1998) An updated table of the best binary codes known. In: Pless V, Huffman W (eds.) Handbook of coding theory. Elsevier, Amsterdam

24. Liu Q, Wang L, Frutos AG, Condon AE, Corn R, Smith LM (2000) Nature 403:175–179

25. MacWilliams FJ, Slone NJA (1985) The theory of error-correcting codes. North-Holland, Amsterdam

26. Marathe A, Condon A, Corn R (2001) On combinatorial DNA word design. J Comput Biol 8:201–220

27. Milenkovic O, Kashyap N (2006) On the design of codes for DNA computing. LNCS 3969:100–119

28. Rykov V, Macula A, Torney D, White P (2001) DNA sequences and quaternary cyclic codes. Proc ISIT 2001, Washington, DC

29. Schulman L, Zuckerman D (1999) Asymptotic good codes correcting insersions, deletions, and transpositions. IEEE Trans Inform Theory 45:2552–2557

30. Sloane L (1999) On single-deletion-correcting codes. In: Arasu K, Seress A (eds.) Codes and Designs. Walter de Gruyter, Berlin

31. Tenengolts G (1984) Nonbinary codes, correcting single deletion or insertion. IEEE Trans Inform Theory 30:766–769

32. Tulpan D, Hoos H, Condon A (2002) Stochastic local search algorithms for DNA word design. LNCS 2568:229–241

33. Tulpan D, Hoos H (2003) Hybrid randomized neighborhood improve stochastic local search for DNA code design. LNAI 2671:418–433

Chapter 5
Non-Autonomous DNA Models

Abstract Early biomolecular computing research focussed on laboratory-scale human-operated DNA models of computation for solving complex computational problems. These models generate large combinatorial libraries of DNA to provide search spaces for parallel filtering algorithms. Many different methods for library generation, solution filtering, and output generation were experimentally studied. This chapter addresses the basic filtering models and describes two basic computationally complete and universal DNA models of computation, splicing model and sticker model.

5.1 Seminal Work

Adleman's first experiment and Lipton's first theoretical work can be considered as the seminal work in DNA computing.

5.1.1 Adleman's First Experiment

The idea of performing massively parallel computations in nanotechnology was first stated by R. Feynman in the late 1950s. In 1994, L. Adleman was the first to demonstrate by a DNA experiment that biomolecular computations are feasible. In this seminal experiment, Adleman solved a small instance of the Hamiltonian path problem. For this, DNA molecules are used as a medium for information storage and this information is manipulated by standard biotechnological operations.

Adleman's first experiment will be briefly recalled. For this, let G be a directed graph. In particular, Adleman employed the graph in Figure 5.1. This graph contains exactly one Hamiltonian path given by $(v_0, v_1, v_2, v_3, v_4, v_5, v_6)$. He aimed to find Hamiltonian paths in G with initial vertex v_0 and final vertex v_6. In Adleman's approach, each vertex v_i is

Z. Ignatova et al., *DNA Computing Models*,
DOI: 10.1007/978-0-387-73637-2_5, © Springer Science+Business Media, LLC 2008

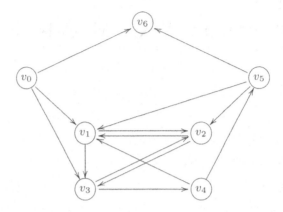

Fig. 5.1 Adleman's graph.

encoded by a single-stranded DNA molecule of 20 nt. Each edge e_{ij} connecting vertex v_i with vertex v_j is encoded by a single-stranded DNA molecule that consists of the complement of the second 3' half-mer of the DNA strand of v_i and the complement of the first 5' half-mer of the DNA strand of v_j. For instance, if the vertices v_0 and v_1 are respectively encoded as

$$5' - \text{AGAGACAG} - 3' \quad \text{and} \quad 5' - \text{ATTCTTTT} - 3'$$

then the edge $e = v_0 v_1$ is encoded as

$$3' - \text{TGTCTAAG} - 5' \, .$$

Thus, hybridization and ligation yields the partially double-stranded molecule indicating the path (v_0, v_1),

$$5' - \text{AGAGACAGATTCTTTT} - 3'$$
$$3' - \text{TGTCTAAG} - 5'$$

Adleman's algorithm can be summarized as follows:

1. Generate random paths in the graph.
2. Keep only those paths that begin with the initial vertex $v_s = v_0$ and end with the final vertex $v_e = v_6$.
3. Retain only those paths that hold all vertices in the graph.
4. Keep only those paths that contain all vertices in the graph at least once.
5. Read out Hamiltonian paths (if any).

The first step provides double-stranded DNA molecules encoding random paths in the graph (Fig. 5.2). This is achieved by annealing and ligation. The second step is implemented by amplifying the product of the first step via PCR. Here, only those single-stranded DNA molecules encoding paths that begin with v_s and end with v_e are amplified. The third step needs length

Fig. 5.2 Encoding of a Hamiltonian path in Adleman's graph.

separation of DNA molecules and is implemented by gel electrophoresis. The fourth step is accomplished by affinity purification. This process permits single-stranded DNA molecules containing a specific subsequence encoding a vertex v_i of the graph to be filtered out from a pool of strands. For this, strands complementary to the strand v_i are synthesized and attached to magnetic beads. The pool of strands is then passed over the beads. Those strands containing the substrand v_i anneal to the beaded complementary strands and can be retained, while the other strands are washed out. Affinity purification is iteratively performed for each vertex of G. The single-stranded DNA molecules surviving this step will encode Hamiltonian paths in G. The last step is implemented by graduated PCR. For this, primer strands v_0 and v_i are taken and the path molecules between these primer strands are amplified. Then the product is length separated by gel electrophoresis to determine the position of the vertex v_i on the path. This process is repeated for each vertex v_i, $1 \leq i \leq 6$. This procedure was successfully implemented in vitro by Adleman, requiring seven days of laboratory work.

Adleman's first experiment can be formally described by a memory-less filtering model. This model is based on the data structure of test tubes. A *test tube* is a finite multiset of strings over a finite alphabet, preferably the DNA alphabet. The operations used by Adleman are the following:

- *prefix-extract* (T, x): Take a test tube T and a string x, and create a test tube that contains all strings in T that have the string x as a prefix.
- *postfix-extract* (T, x): Consider a test tube T and a string x, and generate a test tube that comprises all strings in T that have the string x as a postfix.
- *substring-extract* (T, x): Pick a test tube T and a string x, and provide a test tube that holds all strings in T that have the string x as a substring.
- *length-separate* (T, m): Start with a test tube T and a positive integer m, and produce a test tube that contains all strings in T which have length $\leq m$.
- *detect* (T): Pick a test tube T and output "yes" if T contains at least one DNA molecule; otherwise, output "no".

All extract operations select strings and thus may require the amplification of the resulting test tubes by PCR. A *computation* comprises a finite sequence of these operations. A computation starts with an *initial test tube* and ends with one or more *final test tubes*. Several test tubes may exist simultaneously during a computation.

Example 5.1. Given a graph $G = (V, E)$ with vertex set $V = \{v_1, \ldots, v_n\}$. Take an initial test tube T consisting of a multiset of strings $v_{i_1} \ldots v_{i_k}$, $1 \le i_1, \ldots, i_k \le n$, $k \ge 1$, over the alphabet V, which provide paths in G. Assume that each vertex v_i is encoded by a DNA strand of length m nt. The algorithm ADLEMANHAMILTONIANPATHS filters out the Hamiltonian paths in G that start with v_1 and end with v_n. $\diamond$

Algorithm 5.1 ADLEMANHAMILTONIANPATHS(T, G)

Input: input tube T, graph G
1: $T \leftarrow$ prefix-extract (T, v_1)
2: $T \leftarrow$ postfix-extract (T, v_n)
3: $T \leftarrow$ length-separate $(T, m \cdot n)$
4: **for** $i \leftarrow 2$ to $n - 1$ **do**
5:      $T \leftarrow$ substring-extract (T, v_i)
6: **end for**
7: **return** detect (T)

5.1.2 Lipton's First Paper

The results of L. Adleman was first considered by R. Lipton (1995), providing DNA algorithms for the satisfiability problem and other NP-complete problems. For this, let F be a Boolean expression in n variables given in CNF such as

$$F = (\overline{x}_1 + x_2)(x_1 + \overline{x}_2).$$

Lipton suggested designing for each variable x_i two single-stranded DNA molecules of length 20 nt, one representing "true", x_i^T, and the other representing "false", x_i^F. Using this encoding, Lipton's DNA model employs the following operations:

- *extract* (T, i, b): Consider a test tube T, an integer i, and a value $b \in \{0, 1\}$, and produce a new test tube containing all strings in T for which the ith entry is equal to x_i^T (x_i^F) if $b = 1$ ($b = 0$). The remainder of this extraction is denoted by $\mathrm{rem}(T, i, b)$.
- *merge* $(T_1, \ldots, T_n)$: Take test tubes $T_1, \ldots, T_n$, produce their union $T_1 \cup \ldots \cup T_n$, and place the result into the (possibly empty) tube T_n.

Lipton's algorithm takes an initial test tube T consisting of a multiset of strings $a_1 \ldots a_n$, where a_i encodes a truth value of the ith variable, that is, $a_i = x_i^T$ or $a_i = x_i^F$. The test tube T is formed in the same way that L. Adleman designed the initial test tube for the paths to find Hamiltonian paths in a graph. Here, a specific graph G_n is used for Boolean formulas with n variables (Fig. 5.3).

The algorithm iteratively extracts those DNA strands that satisfy the disjunctive clauses. For instance, in terms of the disjunctive clause $x_1 + \overline{x}_2$, DNA

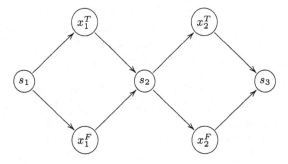

Fig. 5.3 The graph G_2.

strands encoding x_1^T and x_2^F are selected, while the remaining DNA strands are discarded. The final test tube contains only those truth assignments that satisfy all disjunctive clauses and thus the Boolean expression.

Algorithm 5.2 SATISFIABILITYLIPTON(T, F)

Input: input tube T, CNF F in variables $x_1, \ldots, x_n$, test tube T_c initially empty
1: **for** each disjunctive clause c in F **do**
2:     **for** each literal a in c **do**
3:         **if** $a = x_i$ **then**
4:             $T' \leftarrow$ extract $(T, i, 1)$
5:             $T \leftarrow \text{rem}(T, i, 1)$
6:             merge (T', T_c)
7:         **else**
8:             $T' \leftarrow$ extract $(T, i, 0)$
9:             $T \leftarrow \text{rem}(T, i, 0)$
10:            merge (T', T_c)
11:         **end if**
12:     **end for**
13:     clear (T)
14:     $T \leftarrow T_c$
15: **end for**
16: **return** detect (T)

5.2 Filtering Models

This section addresses several basic filtering models of DNA computing.

5.2.1 Memory-Less Filtering

First, a basic filtering model developed by L. Adleman (1996) will be discussed, which is *memory-less* in the sense that the strings themselves do not change during a computation. This model is based on the following operation:

- *separate* (T, T^+, T^-, x): Consider a test tube T and a string x, and produce two new tubes, T^+ and T^-, where T^+ consists of all strings in T which contain the string x as a subsequence, while T^- embraces all strings in T which do not hold the string x as a subsequence.

Example 5.2. Given a graph $G = (V, E)$ with vertex set $V = \{v_1, \ldots, v_n\}$ and edge set $E = \{e_1, \ldots, e_m\}$ so that $e_i = v_{i_1} v_{i_2}$, $1 \le i \le m$. The input test tube T is a multiset of strings $a_1 \ldots a_n$ over the alphabet $\Sigma = \{b_i, g_i, r_i \mid 1 \le i \le n\}$, so that the vertex v_i is assigned the color a_i, $1 \le i \le n$. The algorithm 3-COLORINGS computes the 3-colorings of G available in the initial test tube. This is achieved by singling out all strings that provide edges whose end vertices are equally colored.   ◇

Algorithm 5.3 3-COLORINGS(T, G)

Input: input tube T, graph G
1:   $T' \leftarrow T$
2: **for** $i \leftarrow 1$ to m **do**
3:     separate $(T', T_r, T_{b,g}, r_{i_1})$
4:     separate $(T_{b,g}, T_b, T_g, b_{i_1})$
5:     separate $(T_r, T_r^+, T_r^-, r_{i_2})$
6:     separate $(T_b, T_b^+, T_b^-, b_{i_2})$
7:     separate $(T_g, T_g^+, T_g^-, g_{i_2})$
8:     merge (T_r^-, T_b^-, T_g^-)
9:     $T' \leftarrow T_g^-$
10: **end for**
11: **return** detect (T')

5.2.2 Memory-Based Filtering

L. Adleman (1996) extended the memory-less model to a memory-based filtering model by adding the so-called *flip* operation. For this, take the alphabet $\Sigma = \{b_1, \ldots, b_n\} \cup \{c_1, \ldots, c_n\}$. The flip operation is defined as follows:

- *flip* (T, b_i): Take a test tube T and a character $b_i \in \Sigma$, and produce a new test tube $T^{(b_i)} = \{x^{(b_i)} \mid x \in T\}$, where the string $x^{(b_i)}$ coincides with x up to the ith position, at which it contains b_i provided that x holds c_i.
- *flip* (T, c_i): Pick a test tube T and a character $c_i \in \Sigma$, and create a new test tube $T^{(c_i)} = \{x^{(c_i)} \mid x \in T\}$, where the string $x^{(c_i)}$ coincides with x up to the ith position, at which it holds c_i provided that x contains b_i.

The flip operation switches b_i to c_i (c_i to b_i) in all strings that contain b_i (c_i). For instance, if $x = b_1b_2b_3$, then $x^{(b_2)} = x$ and $x^{(c_2)} = b_1c_2b_3$.

This extended model allows the implementation of n-bit registers. For this, notice that each string $a_1 \ldots a_n$ with $a_i \in \{b_i, c_i\}$, $1 \le i \le n$, can be interpreted as an n-bit register, where the ith location contains 1 if $a_i = b_i$, and 0 if $a_i = c_i$.

Example 5.3. Given an initial test tube T consisting of a multiset of strings $a_1 \ldots a_n$ with $a_i \in \{b_i, c_i\}$, $1 \le i \le n$. The algorithm CONDSETREG sets in each string the kth bit to 1, provided that the ith bit equals 0 and the jth bit equals 1.   ◇

Algorithm 5.4 CONDSETREG(T, i, j, k)

Input: input tube T, positive integers i, j, k
1: separate (T, T^+, T^-, c_i)
2: separate $(T^+, T^{++}, T^{+-}, b_j)$
3: $T' \leftarrow$ flip (T^{++}, b_k)
4: merge (T^-, T^{+-}, T')
5: **return** T'

5.2.3 Mark-and-Destroy Filtering

A filtering model based on the "mark and destroy" paradigm was first proposed by D. Hodgson and coworkers (1996). This model is memory-less and consists of the merge operation plus two further operations:

- *remove* $(T, \{x_1, \ldots, x_m\})$: Take a test tube T and a set of strings $x_1, \ldots, x_m$ and remove any string in T that contains at least one occurrence of the string x_i as a substring, $1 \le i \le m$.
- *copy* $(T, \{T_1, \ldots, T_m\})$: Start with a test tube T and produce a number of m copies T_i of T.

Example 5.4. Take an initial test tube T consisting of a multiset of strings $p_1i_1p_2i_2 \ldots p_ni_n$, where p_j encodes the jth position and i_j is a number from the set $\{1, \ldots, n\}$, $1 \le j \le n$. The algorithm PERMUTATIONS uses the test tube T to provide all n-*permutations of* n (i.e., all strings which contain each number from the set $\{1 \ldots, n\}$ exactly once). For instance, 3-permutations of 3 are $p_11p_22p_33$ and $p_12p_23p_31$. Clearly, the number of n-permutations of n is $n!$ (i.e., n factorial). The remove operation delivers all strings that hold i at the jth position but not at any subsequent position $k > j$.   ◇

A variant of the parallel filtering model was used by Q. Ouyang and coworkers (1997) to tackle the maximum clique problem.

Algorithm 5.5 PERMUTATIONS(T, n)

Input: input tube T, positive integer n
1: **for** $j \leftarrow 1$ to $n - 1$ **do**
2:    copy $(T, \{T_1, \ldots, T_n\})$
3:    **for** $i \leftarrow 1$ to n **do**
4:       **for** $k \leftarrow j + 1$ to n **do**
5:          remove $(T_i, \{p_j 1, \ldots, p_j i - 1, p_j i + 1, \ldots, p_j n, p_k i\})$
6:       **end for**
7:    **end for**
8:    merge $(T_1, \ldots, T_n, T)$
9: **end for**
10: **return** T

Example 5.5. Let $G = (V, E)$ be a graph with vertex set $V = \{v_1, \ldots, v_n\}$ and let $G' = (V, E')$ be the associated complementary graph. Let $E' = \{e_1, \ldots, e_m\}$ be the edge set of G' and write $e_i = v_{i_1} v_{i_2}$, $1 \leq i \leq m$.

Take an initial test tube T consisting of a multiset of strings $p_1 i_1 \ldots p_n i_n$, where p_j encodes the jth position and i_j stands for 0 or 1, $1 \leq j \leq n$. Each such string provides a subset of V consisting of all vertices v_j for which $i_j = 1$, $1 \leq j \leq n$. The algorithm MAXIMUMCLIQUES uses the test tube T to filter out all maximum cliques in G. For this, it iteratively removes all strings that correspond to subsets of V in which two vertices are adjacent in the complementary graph G'. Notice that two vertices adjacent in the complementary graph G' are not adjacent in the original graph G, and vice versa. Thus, two vertices adjacent in the complementary graph G' cannot be members of the same clique. Hence, after the first loop, the only strings left in test tube T correspond to cliques in G. Suppose that each substring p_j has length l nt, and the substrings $i_j = 0$ and $i_j = 1$ have lengths l_0 nt and l_1 nt, respectively. Then the string $p_1 i_1 p_2 i_2 \ldots p_n i_n$ with k substrings $i_j = 1$ has length $nl + kl_1 + (n - k)l_0$ nt. Assume that $l_1 < l_0$. Then the larger the clique in G the smaller the corresponding string, and vice versa. Hence, the maximum cliques can be filtered out by length separation. A variant of this algorithm was implemented by Qi Ouyang et al. in the laboratory. $\diamond$

Algorithm 5.6 MAXIMUMCLIQUES(T, G)

Input: input tube T, graph G
1: **for** $i \leftarrow 1$ to m **do**
2:    remove $(T, \{p_{i_1} 1\})$
3:    remove $(T, \{p_{i_2} 1\})$
4: **end for**
5: **for** $k \leftarrow 1$ to n **do**
6:    **if** detect (length-separate $(T, nl + kl_1 + (n - k)l_0)$) **then**
7:       **return** k
8:    **end if**
9: **end for**

1	2	3
4	5	6
7	8	9

Fig. 5.4 Labelling of chess board.

5.2.4 Split-and-Merge Filtering

A filtering model based on the "split and merge" paradigm was first described by L. Landweber and coworkers (2000). It uses RNA as a computational substrate and addresses a variant of the satisfiability problem, the so-called *knight problem*. The knight problem looks for configurations of knights on an $n \times n$ chess board so that no knight is attacking any other.

Consider the knight problem on a 3×3 chess board (Fig. 5.4). The valid configurations on the board are described by the following Boolean expression:

$$F = (\overline{x}_1 + \overline{x}_6\overline{x}_8)(\overline{x}_2 + \overline{x}_7\overline{x}_9)(\overline{x}_3 + \overline{x}_4\overline{x}_8)(\overline{x}_4 + \overline{x}_3\overline{x}_9)$$
$$(\overline{x}_6 + \overline{x}_1\overline{x}_7)(\overline{x}_7 + \overline{x}_2\overline{x}_6)(\overline{x}_8 + \overline{x}_1\overline{x}_3)(\overline{x}_9 + \overline{x}_2\overline{x}_4) . \tag{5.1}$$

For instance, the expression $\overline{x}_1 + \overline{x}_6\overline{x}_8$ provides the valid configurations for square 1. That is, there is no knight at square 1 (i.e., $x_1 = 0$), or there are no knights at squares 6 and 8 (i.e., $x_6 = x_8 = 0$). The number of squares to be tested for valid configurations can be reduced as the Boolean expression (5.1) is equivalent to the following expression

$$F = (\overline{x}_1 + \overline{x}_6\overline{x}_8)(\overline{x}_2 + \overline{x}_7\overline{x}_9)(\overline{x}_3 + \overline{x}_4\overline{x}_8)(\overline{x}_4 + \overline{x}_3\overline{x}_9)(\overline{x}_6 + \overline{x}_1\overline{x}_7) . \tag{5.2}$$

The algorithm KNIGHTPROBLEM provides a biomolecular algorithm for the knight problem. To this end, design for each variable x_i two single-stranded DNA molecules of length l nt, one representing "true", x_i^T, and the other representing "false", x_i^F. The initial test tube T consists of a multiset of RNA strings $a_1 \ldots a_m$, where m is the number of squares and a_i encodes whether a knight is present at the ith square, $a_i = x_i^T$, or not, $a_i = x_i^F$ (Fig. 5.5). The test tube is constructed from an appropriate DNA test tube by in vitro transcription. Landweber and coworkers prepared the DNA test tube by "mix and split" phosphoramidite chemistry.

Algorithm 5.7 KNIGHTPROBLEM(T, n)

Input: input tube T, $n \times n$ square
 1: **for** each square i **do**
 2:    split (T, T_1, T_2)
 3:    remove $(T_1, \{x_i^T\})$
 4:    remove $(T_2, \{x_i^F\})$
 5:    **for** each square j attacking square i **do**
 6:        remove $(T_2, \{x_j^T\})$
 7:    **end for**
 8:    merge (T_1, T_2, T)
 9: **end for**
10: **return** detect (T)

For each square i, the test tube is equally split into two tubes T_1 and T_2. The subsequent removal operations remove from test tube T_1 all strands that have a knight at square i. Moreover, in test tube T_2 all strands are removed that have no knight at square i as well as strands that provide a knight in an attacking position. Thus, the merging of both test tubes provides valid configurations for the ith square.

The removal operation makes use of the enzyme Ribonuclease H (RNase H). For instance, in order to remove all strands that have a knight at square i, DNA strands which are complementary to the substrand x_i^T are added to the solution. These DNA strands anneal to the RNA strands containing the substrand x_i^T and in this way form DNA/RNA duplexes. Then RNase H is added to the solution. This enzyme digests DNA/RNA duplexes by cleaving the 3'-O-P bond of RNA to produce 3'-hydroxyl and 5'-phosphate terminated products, while it leaves intact those RNA strands with the substrand x_i^F.

The experiment conducted by Landweber et al. for the 3×3 knight problem was encouraging in the sense that 43 sampled output strands contained only one illegal configuration.

A similar "split and merge" approach was used by L. Adleman and coworkers (2002) to tackle a 3-SAT problem with 20 variables in a DNA based experiment. For this, let F be a Boolean expression in n variables given in CNF such as

$$F = (x_2 + \overline{x}_3)(x_1 + \overline{x}_2 + x_3)(\overline{x}_1 + x_2).$$

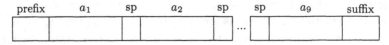

Fig. 5.5 Encoding of strands in the initial test tube for the 3×3 knight problem. The prefix and suffix regions facilitate PCR. Each literal is represented by a DNA strand of 15 nt and two adjacent literals are separated by spacers (sp) of 5 nt.

For each variable x_i, two single-stranded DNA molecules of length l nt are designed, one representing "true", x_i^T, and the other respresenting "false", x_i^F.

The algorithm 3-SATISFIABILITY takes an initial test tube T consisting of a multiset of strings $a_1 \ldots a_n$, where a_i encodes whether the ith variable is assigned the value "true", $a_i = x_i^T$, or "false", $a_i = x_i^F$. Thus each truth assignment of the Boolean expression is encoded by a DNA strand of length ln nt. But a Boolean expression in CNF evaluates to "true" if and only if each disjunctive clause evaluates to "true". Therefore, the algorithm iteratively removes those DNA strands that do not satisfy the disjunctive clauses. For instance, in terms of the disjunctive clause $x_1 + \overline{x}_2 + x_3$, DNA strands encoding x_1^T, x_2^F, and x_3^T are retained, while DNA strands encoding x_1^F, x_2^T, or x_3^F are discarded. This can be achieved by adding DNA strands to the solution that are complementary to x_1^F, x_2^T, or x_3^F. These strands bind to DNA strands that are falsified by the disjunctive clause and thus are partially double-stranded. Such strands can be separated from the single DNA strands by length separation. Hence, the final test tube contains only those truth assignments that are satisfied by the Boolean expression.

In the experiment conducted by Adleman et al, the truth assignments of the variables, x_i^T and x_i^F, were encoded by DNA strands of 15 nt, and thus each DNA strand in the initial test tube was 300 nt long. The chosen Boolean expression (3-SAT) had a unique truth assignment and was indeed detected in the final test tube. This experiment is the largest problem instance successfully solved by a DNA experiment to date.

Algorithm 5.8 3-SATISFIABILITY(T, F)

Input: input tube T, CNF F in variables $x_1, \ldots, x_n$
1: **for** each disjunctive clause c in F **do**
2:    **for** each literal a in c **do**
3:       **if** $a = x_i$ **then**
4:          remove (T, x_i^F)
5:       **else**
6:          remove (T, x_i^T)
7:       **end if**
8:    **end for**
9: **end for**
10: **return** detect (T)

5.2.5 Filtering by Blocking

A filtering model based on the paradigm of blocking was first described by G. Rozenberg and coworkers (2003). A blocking algorithm starts with an initial test tube of potential solutions given by single-stranded DNA molecules, to which a set of complementary falsifying DNA (blockers) is added. A library molecule not representing a solution will hybridize with a blocker to form a perfect double-stranded DNA molecule, while a library molecule corresponding to a solution will remain single-stranded or form a double-stranded molecule with mismatched base pairs, depending on encoding and experimental conditions.

Consider an instance of the satisfiability problem. For this, let F be a Boolean expression in n variables given in CNF such as

$$F = (\overline{x}_1 + x_2 + \overline{x}_3)(x_1 + \overline{x}_2 + x_4)(\overline{x}_1 + x_3 + \overline{x}_4)(x_2 + x_3 + \overline{x}_4) .$$

Each truth assignment of the variables gives rise to a string $x = a_1 \ldots a_n$ over the alphabet $\Sigma = \{0, 1\}$, where the value of $a_i = 1$ ($a_i = 0$) assigns "true" ("false") to the variable x_i, $1 \leq i \leq n$. The truth values are encoded by single DNA strands. In this way, each truth assignment is encoded by a single DNA strand, too. For instance, if "true" and "false" are given by the respective sequences 5'-GTCTGA-3' and 5'-ATCACC-3', the truth assignment 1010 is represented by the single DNA strand (with vertical bars indicating the substrands)

$$5' - \text{GTCTGA} | \text{ATCACC} | \text{GTCTGA} | \text{ATCACC} - 3' .$$

A Boolean expression in CNF is falsified if and only if one clause is falsified. Therefore, the blockers are given by the single DNA strands complementary to truth assignments falsifying a clause. For instance, the truth assignment 1010 falsifies the first clause in the above CNF and thus gives rise to the blocker

$$3' - \text{CAGACT} | \text{TAGTGG} | \text{CAGACT} | \text{TAGTGG} - 5' .$$

An experimental challenge in implementing blocking algorithms is to separate perfectly matched molecules from partially mismatched ones. One promising approach is based on PCR inhibition. For this, the blockers are made of peptide nucleic acid (PNA), chemically similar to DNA or RNA but not naturally occurring in any living species. PNA has a backbone of N-(2-aminoethyl)-glycine units linked by peptide bonds, and the purine and pyrimidine bases are linked to the backbone by methylene carbonyl bonds. The melting temperature of PNA/DNA complexes is eventually higher than that of DNA/DNA double helices. Thus, by carefully controlling the temperature in PCR, falsified molecules given by perfect double-stranded PNA/DNA can be made unavailable for DNA polymerase, while the remaining molecules are selectively amplified. Another approach is based on mutation detection using the enzyme CEL I, which leaves perfect double-stranded DNA untouched and cleaves the remaining DNA. While experimental data for the first approach are lacking, the second approach has shown promising results.

5.2.6 Surface-Based Filtering

Finally, two filtering models are addressed that are based on surface DNA chemistry. The model of L. Smith and coworkers (2000) makes use of the "mark and destroy" paradigm and was employed to solve an instance of the satisfiability problem. In this model, single-stranded DNA molecules are attached to a surface such as glass, gold or silicon, and the DNA molecules are subject to the following operations:

- *mark* (T, B): Start with a test tube T and a Boolean expression B and mark all strands in T which satisfy B.
- *unmark* (T): Consider a test tube T and unmark all marked strands in T.
- *delete* (T, C): Take a test tube T and a condition C, marked or unmarked, and remove all strands in T according to C.

The algorithm SATSURFACE1 provides a DNA algorithm for solving the satisfiability problem. For this, let F be a Boolean expression in n variables given in CNF such as

$$F = (x_2 + \overline{x}_3)(x_1 + \overline{x}_2 + x_3)(\overline{x}_1 + x_2) .$$

For each of the 2^n combinations of truth values, a multitude of single-stranded DNA molecules is designed and attached to the surface. In this way, each solution of the SAT problem is represented as an individual set of affixed strands. In the example, there are eight such strands as illustrated in Table 5.1.

For each disjunctive clause c in the expression F, the truth combinations that satisfy the clause c are marked. For instance, in terms of the clause $x_2 + \overline{x}_3$, the strands s_1, s_3, s_4, s_5, s_7, and s_8 are marked, while the strands s_2 and s_6 are not. After this, the unmarked strands are deleted and the marked strands are unmarked. As the expression F is in CNF, it evaluates to "true" if and only if the final test tube is non-empty.

Algorithm 5.9 SATSURFACE1(T, F)

Input: input tube T, CNF F
1: **for** each disjunctive clause c in F **do**
2:     mark (T, c)
3:     delete $(T, \text{unmarked})$
4:     unmark (T)
5: **end for**
6: **return** detect (T)

The mark operation is implemented by adding to the test tube an excess of strands that are Watson-Crick complementary to the satisfied single strands. These satisfied strands will form double-stranded molecules and thus will be subsequently protected from deletion. The delete operation employs an

Table 5.1 Encoding of truth assignment in SATSURFACE1.

Strand	Sequence	$x_1 x_2 x_3$
s_1	$5' - \text{CAACAACAA} - 3'$	000
s_2	$5' - \text{TAATAATAA} - 3'$	001
s_3	$5' - \text{AGGAGGAGG} - 3'$	010
s_4	$5' - \text{CGGCGGCGG} - 3'$	011
s_5	$5' - \text{CAAAGAAAC} - 3'$	100
s_6	$5' - \text{TGGGTGGGT} - 3'$	101
s_7	$5' - \text{TCCCTCCCT} - 3'$	110
s_8	$5' - \text{AAATAAATA} - 3'$	111

E. coli exonuclease I to cleave phosphodiester bonds in the 3' to 5' direction of the remaining single-stranded DNA molecules. In the example, the single-stranded molecules corresponding to the strands s_2 and s_6 are digested. Finally, the unmark operation denaturates the double-stranded molecules and filters out the strands added in the first step. This leaves only immobilized single-stranded molecules which satisfy the clause.

Another surface-based filtering model devised by Y. Sakakibara and A. Suyama (2000) can be used to design so-called *universal DNA chips*. For this, let F be a Boolean expression in n variables given in *disjunctive normal form (DNF)* (i.e., F is a disjunction of conjunctions (clauses) of literals), such as

$$F = x_1 \overline{x}_2 + \overline{x}_3 x_4 .$$

Each variable is encoded by a single-stranded DNA molecule of fixed length and its negation is encoded by the corresponding Watson-Crick complement. Each conjunctive clause of F is encoded by the sequence of occurring literals, and this strand is prefixed by a marker sequence. The whole Boolean expression F is encoded by the sequence of conjunctive clauses, and the clauses are separated by stopper sequences (Fig. 5.6). These DNA molecules can be affixed to a surface and are subject to the following operation:

- *mark* (T, F, b): Pick a test tube T, a Boolean expression F in DNF, a truth assignment b, and add an excess of encoded DNA of the literals to the test tube: If $b_i = 1$ then add encoded DNA of x_i, and if $b_i = 0$ then add encoded DNA of the Watson-Crick complement of x_i.

The algorithm SATSURFACE2 provides a DNA algorithm for solving the satisfiability problem. Notice that a conjunctive clause is satisfied by the truth assignment if and only if the mark operation leaves the correspond-

marker	x_1	$\overline{x}_2$	stopper	marker	$\overline{x}_3$	x_4

Fig. 5.6 Encoding of DNF expression F.

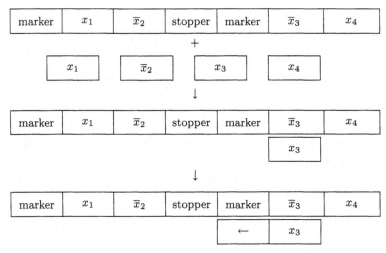

Fig. 5.7 DNA computation for Boolean expression F with truth assignment $b = 1011$.

ing substrand single-stranded. Thus, if a conjunctive clause is not satisfied, then the associated substrand is (partially) double-stranded. Then PCR will extend this substrand and this extension will stop at the stopper sequence. Consequently, the Boolean expression is satisfied by the truth assignment if and only if at least one marker sequence is not double-stranded (Fig. 5.7). These marker strands can be detected by adding complementary fluorescently tagged markers to the solution. Then the intensity of fluorescence becomes proportional to the satisfiability level of the Boolean expression.

Algorithm 5.10 SatSurface2(T, F, b)

Input: input tube T, DNF F, truth assignment b
1: mark (T, F, b)
2: pcr (T)
3: **return** detect (T)

In view of DNA chips, the Boolean variables are encoded by DNA words termed *DNA coded numbers* (DCNs), which exhibit uniform melting temperature and low mishybridization or self-folding potential. An mRNA transcript (or the corresponding complementary DNA) can be converted to a DCN by using template sequences that hybridize with both the mRNA transcript and the DCN. By changing the set of template sequences, the set of mRNAs detected by the chip can be changed. In this sense, the chip is considered to be *universal*.

A universal DNA chip allows the development of an intelligent DNA chip which can process rules for molecular diagnosis such as the following:

If gene A is not expressed and gene B is expressed, then there is a danger of disease C.

The condition in this rule can be represented by a Boolean expression in DNF as described above so that an expressed gene is associated with a DCN and an unexpressed gene corresponds to the Watson-Crick complement of a DCN. An advantage of intelligent DNA chips is that the intensity of fluorescence is proportional to the expression level of the genes and also proportional to the satisfiability level of the condition.

5.3 Sticker Systems

The sticker system model introduced by L. Adleman and coworkers (1996) is one of the most popular non-autonomous DNA models. It belongs to the class of filtering models and can be viewed as an implementation of a register machine.

5.3.1 Sticker Machines

This section describes the basic structure of sticker machines.

Sticker Memory

The sticker system model has a random access memory which requires no strand extension. The memory of the sticker system model consists of so-called memory complexes. A *memory complex* is a DNA strand that is partially double-stranded and can be viewed as an encoding of an n bit number. Each memory complex is formed by two basic types of single-stranded DNA molecules, referred to as memory strands and sticker strands. A *memory strand* is a single-stranded DNA molecule consisting of l nt in length. A memory strand contains n non-overlapping substrands, each of which has m nt so that $l = mn$. As an example, consider the following memory strand for $n = 4$ and $m = 6$ (with vertical bars indicating the substrands):

$$5' - \text{AAAAAA} \,|\, \text{TTTTTT} \,|\, \text{GGGAAA} \,|\, \text{CCCTTT} - 3' \,.$$

Each *sticker strand* is m bases long. As an example, consider the following sticker strands for $m = 6$:

$$3' - \text{AAAAAA} - 5', \qquad 3' - \text{CCCTTT} - 5', \qquad 3' - \text{GGGAAA} - 5'.$$

Each sticker strand is required to be complementary to exactly one of the n substrands in a memory strand. For this, the substrands of a memory strand should differ with respect to several base positions. Each substrand of a memory strand will be identified with one bit position. If a sticker strand is annealed to its matched substrand on a memory strand, the particular substrand is *on*; otherwise, it is *off*. In this way, memory complexes can represent binary numbers, where a substrand being on represents bit 1 and a substrand being off represents bit 0.

Example 5.6. Consider the following memory complexes:

```
5'-AAAAAA|TTTTTT|GGGAAA|CCCTTT-3'
```

```
5'-AAAAAA|TTTTTT|GGGAAA|CCCTTT-3'
    TTTTTT        CCCTTT
```

```
5'-AAAAAA|TTTTTT|GGGAAA|CCCTTT-3'
         AAAAAA|CCCTTT|GGGAAA
```

The encoded 4-bit strings are in turn 0000, 1010, and 0111.   ◊

Sticker Operations

There are basically five operations in the sticker system model:

- *merge* $(T_1, \ldots, T_n)$: Take test tubes $T_1, \ldots, T_n$, produce their union $T_1 \cup \ldots \cup T_n$, and put the result into the (possibly empty) test tube T_n.
- *separate* (T, T^+, T^-, i): Consider a test tube T and an integer i and create two new tubes T^+ and T^-, where T^+ consists of all memory complexes in T in which the ith substrand is on, while T^- is comprised of all memory complexes in T in which the ith substrand is off.
- *set* (T, i): Start with a test tube T and an integer i and generate a test tube in which the ith substrand of each memory complex in T is turned on.
- *clear* (T, i): Pick a test tube T and an integer i and produce a test tube in which the ith substrand of each memory complex in T is turned off.
- *discard* (T): Take a test tube T and empty its contents.

The operation merge is implemented by simply pouring the contents of several test tubes into a new one. However, if DNA is not gently handled, it may be fragmented by shear forces into smaller pieces. Moreover, DNA molecules may remain stuck to the walls of a tube and thus become eventually lost.

The operation separate divides the contents of a test tube into two test tubes depending on the ith substrand. For this, oligonucleotide probes can be designed that are complementary to the ith substrand and are attached to the tube's wall. Then the memory complexes in which the ith substrand is off will probably anneal to such probes. The probes must have lower binding affinity than the sticker strands so that the annealed memory complexes can be recovered from the probes without losing the stickers.

The operation set can be implemented by adding an excess amount of sticker strands to the test tube and letting them hybridize with the memory complexes. In order to remove unused sticker strands, a universal region can be added to each memory strand so that no sticker can anneal to this region.

The operation clear is the most problematic one, since melting does not work because the sticker strands in all memory complexes will fall off. However, this operation can be eliminated without sacrificing the computational completeness of the sticker system model. Indeed, instead of clearing the ith substrand, an unused substrand can be set in an extended memory complex.

Sticker Computations

A *sticker computation* consists of a finite sequence of sticker operations. The input of a sticker computation is a test tube called initial test tube, while the output is a sequence of test tubes called final test tubes. The output of a final test tube is read in the sense that each of its memory complexes is analyzed by isolating the annealed stickers, or it is reported that it contains no memory complexes. Many sticker algorithms require additional test tubes, which are considered to be empty at the beginning of a computation.

Sticker Machine

A *sticker machine* may be thought of as a robotic workstation that consists of some robotics (arms, pumps, heaters, coolers), a microprocessor that controls the robotics, and a series of test tubes: *data tubes* that hold memory complexes, *sticker tubes* each of which containing particular stickers, and *separation operator tubes* that hold probes for a particular substrand. The microprocessor controls the robotics and test tubes so that the operations of a sticker program can be sequentially executed. The *complexity* of a sticker algorithm is counted by the total number of sticker operations. The costs necessary to generate an initial test tube and to read the final test tubes are ignored.

5.3.2 Combinatorial Libraries

Each DNA filtering algorithm uses as input an initial test tube containing candidate solutions. Such a test tube is also termed a *combinatorial library*. The search for candidate solutions during a filtering computation can be reduced by using an initial test tube that contains fewer combinatorial types.

A basic type is the *Pascal library* introduced by K.-H. Zimmermann (2002). For this, let k and n be positive integers. A string $x = a_1 \ldots a_k$ over the alphabet $\{1, \ldots, n\}$ of length k is called a *k-combination* of n if x is *strictly increasing* (i.e., $a_1 < a_2 < \ldots < a_k$). For instance, the 2-combinations of 4 are 12, 13, 14, 23, 24 and 34. The number of k-combinations of n is denoted by $\binom{n}{k}$, called a *binomial number*. In particular, $\binom{n}{k} = 0$, if $k > n$, $\binom{n}{0} = 1$, $\binom{n}{n} = 1$ and $\binom{n}{1} = n$. Moreover, if $1 \leq k \leq n$ then

$$\binom{n}{k} = \binom{n-1}{k-1} + \binom{n-1}{k}. \tag{5.3}$$

Each k-combination $x = a_1 a_2 \ldots a_k$ of n describes a k-subset $f(x) = \{a_1, a_2, \ldots, a_k\}$ of $\{1, \ldots, n\}$. The assignment $f : x \mapsto f(x)$ provides a bijection between the set of all k-combinations of n and the set of all k-subsets of $\{1, \ldots, n\}$. If we denote the set of all k-subsets of $\{1, \ldots, n\}$ by $\binom{n}{k}$, then

$$\binom{n}{k} = \left| \binom{n}{k} \right|. \tag{5.4}$$

Let $n \geq k$ and $m \geq 0$. An $[n + m, \binom{n}{k}]$ *library* is a test tube given by a multiset of DNA molecules encoding as $n + m$ bit numbers all k-combinations of n so that the last m symbols are 0. For instance, a $[7, \binom{4}{3}]$ library is given by a multiset of strings 1110|000, 1101|000, 1011|000, and 0111|000, where vertical bars indicate the substrings.

A combinatorial $[n + m, \binom{n}{k}]$ library can be fabricated inductively by making use of Eq. (5.3). That means, given an $[n + m, \binom{n-1}{k}]$ library and an $[n + m, \binom{n-1}{k-1}]$ library. First, an appropriate number of stickers is added to the latter library for the nth bit and then both libraries are merged to provide an $[n + m, \binom{n}{k}]$ library.

5.3.3 Useful Subroutines

In the following, let $G = (V, E)$ be a graph with vertex set $V = \{v_1, \ldots, v_n\}$ and edge set $E = \{e_1, \ldots, e_m\}$. The ith edge is denoted by $e_i = v_{i_1} v_{i_2}$, $1 \leq i \leq m$. It is assumed that there is a sticker machine whose microprocessor has a description of the graph G. In this way, the graph itself need not be specified by memory complexes. Some useful procedures introduced by K.-H. Zimmermann (2002) and I. Martinéz-Pérez (2007) will be described first, which serve as building blocks for the construction of more complex sticker algorithms. All algorithms presented were validated by the JAVA-based STICKERSIM library written by O. Scharrenberg (2007).

Edge-Induced Graphs

The algorithm EDGEINDUCEDGRAPHS provides all subgraphs of a graph G that are induced from the k-subsets of edges, where $1 \le k \le m$. The input of the algorithm is an $[m + n, \binom{m}{k}]$ library T, providing the encoded DNA of all k-subsets of edges. The algorithm operates in a bit-vertical fashion as it considers in parallel the ith substrands of the memory complexes. For those memory complexes whose ith substrand is on, the edge e_i occurs in the corresponding k-set of edges. If the ith substrand is on, the substrands $m + i_1$ and $m + i_2$ are turned on, indicating the vertices of the edge e_i. In the final test tube, the memory complexes correspond to the subgraphs of G, which are induced from the k-sets of edges. The algorithm requires $4m$ steps.

Example 5.7. In view of the graph in Figure 5.8 and the library T with $k = 2$, the algorithm yields the memory complexes in Table 5.2. $\diamond$

Algorithm 5.11 EDGEINDUCEDGRAPHS(T, m, n)

Input: $[m + n, \binom{m}{k}]$ library T
  1: **for** $i \leftarrow 1$ to m **do**
  2:      separate (T, T^+, T^-, i)
  3:      set $(T^+, m + i_1)$
  4:      set $(T^+, m + i_2)$
  5:      merge (T^+, T^-, T)
  6: **end for**
  7: **return** T

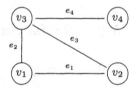

Fig. 5.8 A graph G.

Table 5.2 Final test tube of EDGEINDUCEDGRAPHS.

e_1	e_2	e_3	e_4	v_1	v_2	v_3	v_4
1	1	0	0	1	1	1	0
1	0	1	0	1	1	1	0
1	0	0	1	1	1	1	1
0	1	1	0	1	1	1	0
0	1	0	1	1	0	1	1
0	0	1	1	0	1	1	1

Weightening

The algorithm WEIGHTENING extracts from an input test tube T_0 those memory complexes in which exactly k of the substrands $m+1, \ldots, m+n$ are turned on, where $0 \le k \le n$. At the end of the loop (1–7), the test tube T_i, $0 \le i \le n$, contains all memory complexes in which exactly i of the substrands $m+1, \ldots, m+n$ are turned on. Thus the test tube T_k provides the output of the algorithm. The sticker algorithm requires $2n\frac{n+1}{2} = n^2 + n$ steps. The test tubes $T_{k+1}, \ldots, T_n$ are not required, so that the second statement can be altered as follows:

$$2: \quad \textbf{for } j \leftarrow \min\{i, k\} \text{ down to } 0 \textbf{ do} \,.$$

This algorithm needs $2(1 + 2 + \ldots + k + (n-k)(k+1)) = 2n(k+1) - k^2 - k$ steps.

Example 5.8. Consider an input test tube T_0 providing encoded DNA of the memory complexes 00000, 10101, 01111, and 11010. The computation of WEIGHTENING for $m = 0$ and $n = 5$ is shown in Table 5.3. $\diamond$

Algorithm 5.12 WEIGHTENING(T_0, m, n, k)

Input: input test tube T_0
1: **for** $i \leftarrow 0$ to $n - 1$ **do**
2:     **for** $j \leftarrow i$ down to 0 **do**
3:         separate $(T_j, T^+, T^-, m+i+1)$
4:         merge (T^+, T_{j+1})
5:         merge (T^-, T_j)
6:     **end for**
7: **end for**
8: **return** T_k

Table 5.3 Computation of WEIGHTENING.

	T_0	T_1	T_2	T_3	T_4	T_5
initial	00000					
	10101					
	01111					
	11010					
i=1	00000	10101				
sep. on 1	01111	11010				
i=2	00000	10101	11010			
sep. on 2		01111				
i=3	00000		11010			
			10101			
sep. on 3			01111			
i=4	00000		10101	11010		
sep. on 4				01111		
i=5	00000			11010	01111	
sep. on 5				10101		

Complement

The algorithm COMPLEMENT yields the complements of all k-subsets of vertices in a graph $G = (V, E)$, where $1 \leq k \leq n$. That is, for each k-subset S of V, the algorithm finds the complementary subset $\overline{S}$ in V. The input of the algorithm is a $[2n, \binom{n}{k}]$ library T, providing encoded DNA of all k-subsets of vertices. The algorithm turns on substrand $i + n$ for those memory complexes whose ith substrand is turned off. As a result, the complement of a subset of V given by the first n substrands is composed of the last n substrands. The algorithm requires $3n$ steps.

Example 5.9. Given an initial test tube T with $n = 4$ and $k = 2$, the output of the algorithm COMPLEMENT is illustrated in Table 5.4.   ◇

Algorithm 5.13 COMPLEMENT(T, n)

Input: $[2n, \binom{n}{k}]$ library T
1: **for** $i \leftarrow 1$ to n **do**
2:    separate (T, T^+, T^-, i)
3:    set $(T^-, i + n)$
4:    merge (T^+, T^-, T)
5: **end for**
6: **return** T

Table 5.4 Final test tube of COMPLEMENT.

n_1	n_2	n_3	n_4	c_1	c_2	c_3	c_4
1	1	0	0	0	0	1	1
1	0	1	0	0	1	0	1
1	0	0	1	0	1	1	0
0	1	1	0	1	0	0	1
0	1	0	1	1	0	1	0
0	0	1	1	1	1	0	0

Algorithm 5.14 VERTEXINDUCEDGRAPHS(T, m, n)

Input: $[n + m, \binom{n}{k}]$ library T
1: **for** $i \leftarrow 1$ to m **do**
2:     separate (T, T^+, T^-, i_1)
3:     separate $(T^+, T^{++}, T^{+-}, i_2)$
4:     set $(T^{++}, n + i)$
5:     merge (T^-, T^{+-}, T^{++}, T)
6: **end for**
7: **return** T

Vertex-Induced Graphs

The algorithm VERTEXINDUCEDGRAPHS yields all subgraphs of a graph G that are induced by the k-subsets of vertices, where $1 \leq k \leq n$. The input of the algorithm is an $[n + m, \binom{n}{k}]$ library T, which provides encoded DNA of all k-subsets of vertices. The algorithm loops over all edges in G. In terms of the edge $e_i = v_{i_1} v_{i_2}$, if a k-set has both the i_1th and the i_2th substrands turned on, the corresponding memory complex finds its way into the tube T^{++}, and the strand $n + i$ encoding the edge e_i is turned on. The algorithm requires $4m$ steps.

Example 5.10. Given the graph in Figure 5.8 and the initial test tube T with $k = 3$, the algorithm produces the final test tube shown in Table 5.5.   ◇

Table 5.5 Final test tube of VERTEXINDUCEDGRAPHS.

v_1	v_2	v_3	v_4	e_1	e_2	e_3	e_4
1	1	1	0	1	1	1	0
1	1	0	1	1	0	0	0
1	0	1	1	0	1	0	1
0	1	1	1	0	0	1	1

Algorithm 5.15 SPANNINGTREES(T, m, n)

Input: $[n + m, \binom{m}{n-1})]$ library T
 1: $T \leftarrow$ EdgeInducedGraphs (T, m, n)
 2: **for** $i \leftarrow 1$ to n **do**
 3:     separate $(T, T^+, T^-, m + i)$
 4:     $T \leftarrow T^+$
 5: **end for**
 6: **return** T

Spanning Trees

The algorithm SPANNINGTREES provides all spanning trees of a graph G. It makes use of an $[m + n, \binom{m}{n-1})]$ library T, which provides encoded DNA of all $n - 1$-subsets of edges. This library is used by EDGEINDUCEDGRAPHS to compute all induced subgraphs. If the vertex v_i belongs to such an induced subgraph, the substrand $m+i$ in the corresponding memory complex is turned on. The memory complexes with the substrand $m + i$ turned off are removed step by step from the test tube, leaving at the end the memory complexes in which the substrands $m+1, \ldots, m+n$ are turned on. By Theorem 2.18, these complexes correspond one-to-one to the spanning trees of G. The algorithm requires $4m + n$ steps.

Example 5.11. Given the graph in Figure 5.8, EDGEINDUCEDGRAPHS yields the memory complexes illustrated in Table 5.6. The last three memory complexes correspond to the spanning trees of the graph and end up in the final test tube of SPANNINGTREES (Fig. 5.9). $\diamond$

Incidence Relation

The algorithm INCIDENCERELATION provides the incidence relation between vertices and edges in a graph G. The input of the algorithm is an $[m+n, \binom{n}{k})]$ library T, providing encoded DNA of all k-subsets of vertices, where $1 \leq k \leq n$. Two additional parameters are introduced, a lower (l) and upper (u) bound on the set of substrands of size n, where $1 \leq l < u \leq n$. For those memory

Table 5.6 Test tube after EDGEINDUCEDGRAPHS in SPANNINGTREES.

e_1	e_2	e_3	e_4	v_1	v_2	v_3	v_4
1	1	1	0	1	1	1	0
1	1	0	1	1	1	1	1
1	0	1	1	1	1	1	1
0	1	1	1	1	1	1	1

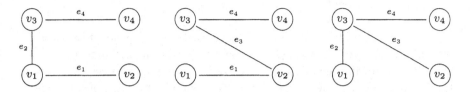

Fig. 5.9 The spanning trees of graph G.

complexes whose ith substrand is turned on, $l \le i \le u$, the algorithm verifies
in parallel if the vertex-edge pair (v_i, e_j) is incident and if so, turns on the
$u+j$th substrand corresponding to the incident edge (statements 4–5). At the
end of the loop, the strands composed of the last m substrands provide the
incidence pairs (v_i, e_j) between vertices and edges. The algorithm requires
$n(m + 2)$ steps.

Algorithm 5.16 INCIDENCERELATION(T, l, u, m, n)

Input: $[m + n, \binom{n}{k}]$ library T, $1 \le l < u \le n$
1: **for** $i \leftarrow l$ to u **do**
2:     separate (T, T^+, T^-, i)
3:     **for** $j \leftarrow 1$ to m **do**
4:         **if** incident (i, j) **then**
5:             set $(T^+, u + j)$
6:         **end if**
7:     **end for**
8:     merge (T^+, T^-, T)
9: **end for**
10: **return** T

Example 5.12. Given the graph G in Figure 5.8 and the library T with $n = u = 4$, $l = 1$, and $k = 2$, the output of the algorithm is given in Table 5.7.
The table indicates that in the first memory complex, the incident edges of
v_1 and v_2 are e_1, e_2, and e_3, because v_1 is incident with e_1 and e_2, and v_2 is
incident with e_1 and e_3. $\diamond$

Table 5.7 Final test tube of INCIDENCERELATION.

v_1	v_2	v_3	v_4	e_1	e_2	e_3	e_4
1	1	0	0	1	1	1	0
1	0	1	0	1	1	1	1
1	0	0	1	1	1	0	1
0	1	1	0	1	1	1	1
0	1	0	1	1	0	1	1
0	0	1	1	0	1	1	1

Independent Subsets

The algorithm INDEPENDENTSUBSETS constructs an independent set in each subgraph with k vertices of a graph. The input of the algorithm is a $[2n, \binom{n}{k}]$ library T, providing encoded DNA of all k-subsets of vertices of G with $1 \leq k \leq n$. Moreover, lower (l) and upper (u) bounds on the set of substrands of size n are given, where $1 \leq l < u \leq n$. The algorithm verifies whether the vertices v_i and v_j are adjacent, $l \leq i < j \leq u$. If so, the jth substrand is cleared and the $(u+1)+(j-l)$th substrand is set (i.e., the vertex v_j is removed from the k-subset and stored in the second set of substrands of size n). At the end of the algorithm, the first n strands of each memory complex provide an independent subset of the initially given k-subset. The algorithm requires $n^2 + 6n$ steps. Notice that the constructed independent subsets depend on the ordering of the vertices.

Algorithm 5.17 INDEPENDENTSUBSETS(T, l, u, n)

Input: $[2n, \binom{n}{k}]$ library T, $1 \leq l < u \leq n$
  1: **for** $i \leftarrow l$ to $u - 1$ **do**
  2:     **for** $j \leftarrow i + 1$ to u **do**
  3:         separate (T, T^+, T^-, i)
  4:         separate (T^+, T^{++}, T^{+-}, j)
  5:         merge (T^-, T^{+-}, T)
  6:         **if** adjacent (v_i, v_j) **then**
  7:            set $(T^{++}, (u + 1) + (j - l))$
  8:            clear (T^{++}, j)
  9:         **end if**
10:         merge (T^{++}, T)
11:     **end for**
12: **end for**
13: **return** T

Example 5.13. Given the graph G in Figure 5.8 and the library T with $n = u = 4$, $l = 1$, and $k = 3$, the algorithm yields the output tube described in Table 5.8. For instance, the second memory complex encodes the independent subset $\{v_1, v_4\}$ of the subgraph spanned by the vertex set $\{v_1, v_2, v_4\}$. $\Diamond$

Table 5.8 Final test tube of INDEPENDENTSUBSETS.

v_1	v_2	v_3	v_4	v_1	v_2	v_3	v_4
1	1	0	0	1	1	1	0
1	0	1	0	1	1	1	1
1	0	0	1	1	1	0	1
0	1	1	0	1	1	1	1
0	1	0	1	1	0	1	1
0	0	1	1	0	1	1	1

Mutually Disjoint Sets

The algorithm MUTUALLYDISJOINTSETS verifies whether the vertex set of a graph G is partitioned into two disjoint subsets. The input of the algorithm is a $[2n, \binom{2n}{k}]$ library T so that the first n substrands provide an encoding of the k-subsets of V. Likewise, the second n substrands give another encoding of subsets of V. The algorithm separates the library T into four tubes T^{++}, T^{+-}, T^{-+}, and T^{--} each of which indicates the presence ($^+$) or absence ($^-$) of vertices v_i and v_{i+n}, respectively. The next statement merges the tubes that contain either vertex v_i or vertex v_{i+n}, while the tubes T^{++} and T^{--} are discarded. This algorithm requires $6n$ steps.

Algorithm 5.18 MUTUALLYDISJOINTSETS(T, n)

Input: $[2n, \binom{2n}{k}]$ library T
1: **for** $i \leftarrow 1$ to n **do**
2:      separate (T, T^+, T^-, i)
3:      separate $(T^+, T^{++}, T^{+-}, i + n)$
4:      separate $(T^-, T^{-+}, T^{--}, i + n)$
5:      merge (T^{+-}, T^{-+}, T)
6:      discard (T^{++})
7:      discard (T^{--})
8: **end for**
9: **return** T

Example 5.14. Assume that an already processed test tube T with $n = 4$ and $k = 2$ exhibits the contents given in Table 5.9. The algorithm eliminates those memory complexes whose ith and $i + n$th substrands have the same value and produces the output tube illustrated in Table 5.10.   ◇

5.3.4 NP-Complete Problems

This section provides a collection of sticker algorithms for NP-complete problems. One distinctive feature of these algorithms is that they solve

Table 5.9 Input test tube of MUTUALLYDISJOINTSETS.

v_1	v_2	v_3	v_4	v_1	v_2	v_3	v_4
1	1	0	0	1	1	0	0
1	0	1	0	0	1	0	1
1	0	0	1	1	0	0	1
0	1	1	0	1	0	0	1
0	1	0	1	0	1	0	1
0	0	1	1	1	1	0	0

Table 5.10 Final test tube of MUTUALLY DISJOINT SETS.

v_1	v_2	v_3	v_4	v_1	v_2	v_3	v_4
1	0	1	0	0	1	0	1
0	1	1	0	1	0	0	1
0	0	1	1	1	1	0	0

NP-complete problems in low polynomial time by shifting complexity into space.

Set Covers

The power of sticker systems was first demonstrated by an algorithm solving the problem of minimum set cover. Let $\{A_1, \ldots, A_m\}$ be a set of subsets of the set $A = \{1, \ldots, n\}$. The algorithm SETCOVERS provides all set covers of A that consist of k elements. The input of the algorithm is an $[m+n, \binom{m}{k}]$ library T which provides encoded DNA of all k-element subsets of $\{A_1, \ldots, A_m\}$. The ith loop inspects the ith subset A_i, and if the ith strand is turned on, then the elements of A_i are stored in the last n substrands. Thus, the set covers are formed by memory complexes in which the last n strands are turned on. These complexes are filtered out by WEIGHTENING. This sticker algorithm needs $m(2 + \sum_i |A_i|) + n^2 + n$ steps. But $\sum_i |A_i| \leq mn$ and thus an upper bound on the number of steps is $m^2 n + n^2 + 2m + n$.

Example 5.15. Consider the subsets $A_1 = \{1, 2\}$, $A_2 = \{2, 3\}$, and $A_3 = \{3, 4\}$ of the set $A = \{1, \ldots, 4\}$, and let $k = 2$. The initial test tube T is illustrated in Table 5.11. After the end of the loop, the tube holds the memory complexes given in Table 5.12 and WEIGHTENING produces the tube in Table 5.13. Hence, the 2-subset $\{A_1, A_3\}$ forms a set cover of A.   ◇

Algorithm 5.19 SETCOVERS(T, m, n, k)

Input: $[m + n, \binom{m}{k}]$ library T
1: **for** $i \leftarrow 1$ to m **do**
2:   separate (T, T^+, T^-, i)
3:   **for** each $j \in A_i$ **do**
4:     set $(T^+, m + j)$
5:   **end for**
6:   merge (T^+, T^-, T)
7: **end for**
8: $T \leftarrow$ Weightening (T, m, n, n)
9: **if** ¬empty (T) **then**
10:   **return** T
11: **else**
12:   report "no k-set cover"
13: **end if**

Table 5.11 Initial test tube of SETCOVERS.

A_1	A_2	A_3	1	2	3	4
1	1	0	0	0	0	0
1	0	1	0	0	0	0
0	1	1	0	0	0	0

Table 5.12 Test tube of SETCOVERS after the loop.

A_1	A_2	A_3	1	2	3	4
1	1	0	1	1	1	0
1	0	1	1	1	1	1
0	1	1	0	1	1	1

Table 5.13 Final test tube of SETCOVERS.

A_1	A_2	A_3	1	2	3	4
1	0	1	1	1	1	1

Vertex Covers

The algorithm VERTEXCOVERS determines whether a graph G exhibits a vertex cover of size k. The input of the algorithm is an $[m + n, \binom{n}{k}]$ library T, providing encoded DNA of all k-subsets of vertices, where $1 \leq k \leq n$. The algorithm first constructs for each k-subset of vertices the set of incident edges by using the procedure INCIDENCERELATION. These edges are stored in the last m substrands of the memory complexes. Those memory complexes in which all m substrands are turned on provide vertex covers. The memory complexes with this property are filtered out by WEIGHTENING. The algorithm requires $m^2 + mn + m + 2n$ steps. The minimum vertex cover problem can be solved by invoking the algorithm for increasing parameters $k = 1, \ldots, n$ (respectively input test tubes) until the corresponding output test tube is non-empty.

Algorithm 5.20 VERTEXCOVERS(T, m, n, k)

Input: $[m + n, \binom{n}{k}]$ library T
1: $T \leftarrow$ IncidenceRelation $(T, 1, n, m, n)$
2: $T \leftarrow$ Weightening (T, n, m, m)
3: **if** $\neg$empty (T) **then**
4:     **return** T
5: **else**
6:     report "no k-vertex cover"
7: **end if**

Table 5.14 Initial test tube of VERTEXCOVERS.

v_1	v_2	v_3	v_4	e_1	e_2	e_3	e_4
1	1	0	0	0	0	0	0
1	0	1	0	0	0	0	0
1	0	0	1	0	0	0	0
0	1	1	0	0	0	0	0
0	1	0	1	0	0	0	0
0	0	1	1	0	0	0	0

Example 5.16. Consider the graph G in Figure 5.8 and take the library T with $k = 2$ containing the memory complexes described in Table 5.14. INCIDENCERELATION produces the tube in Table 5.15, and WEIGHTENING generates the tube in Table 5.16. Thus $\{v_1, v_3\}$ and $\{v_2, v_3\}$ are vertex covers of G.   ◇

Cliques

The algorithm CLIQUES1 provides all cliques of size k in a graph G. The input of the algorithm is an $[n, \binom{n}{k}]$ library T, providing encoded DNA of all k-subsets of vertices, where $1 \leq k \leq n$. The algorithm first calculates the tube T^{++} containing those memory complexes in which the ith and jth substrands are turned on. Then the algorithm checks whether the corresponding vertices are adjacent. If not, the memory complexes are filtered out (statements 6–7). This algorithm requires $n^2 + 5n$ steps.

Table 5.15 Test tube produced by INCIDENCERELATION in VERTEXCOVERS.

v_1	v_2	v_3	v_4	e_1	e_2	e_3	e_4
1	1	0	0	1	1	1	0
1	0	1	0	1	1	1	1
1	0	0	1	1	1	0	1
0	1	1	0	1	1	1	1
0	1	0	1	1	0	1	1
0	0	1	1	0	1	1	1

Table 5.16 Final test tube of VERTEXCOVERS.

v_1	v_2	v_3	v_4	e_1	e_2	e_3	e_4
1	0	1	0	1	1	1	1
0	1	1	0	1	1	1	1

Algorithm 5.21 CLIQUES1(T, n, k)

Input: $[n, \binom{n}{k}]$ library T
 1: **for** $i \leftarrow 1$ to $n - 1$ **do**
 2:    **for** $j \leftarrow i + 1$ to n **do**
 3:       separate (T, T^+, T^-, i)
 4:       separate (T^+, T^{++}, T^{+-}, j)
 5:       merge (T^-, T^{+-}, T)
 6:       **if** adjacent (v_i, v_j) **then**
 7:          merge (T^{++}, T)
 8:       **end if**
 9:       discard (T^{++})
10:    **end for**
11: **end for**
12: **if** $\neg$empty (T) **then**
13:    **return** T
14: **else**
15:    report "no k-clique"
16: **end if**

The maximum clique problem can be solved by invoking CLIQUES1 for decreasing parameters $k = n, n - 1, \ldots$ (respectively input test tubes) until the corresponding output test tube is non-empty.

Example 5.17. Consider the graph G in Figure 5.8 and the initial library T with $k = 3$ providing the memory complexes in Table 5.17. The final tube holds the memory strand in Table 5.18. Thus, the graph G contains a single 3-clique, given by $\{v_1, v_2, v_3\}$. $\Diamond$

Here is another sticker algorithm for the clique problem. The input of the algorithm is an $[m + n, \binom{m}{k}]$ library T, providing encoded DNA of all k-subsets of edges, where $\binom{k}{2} \leq m$. The algorithm determines which of these

Table 5.17 Initial test tube of CLIQUES1.

v_1	v_2	v_3	v_4
1	1	1	0
1	1	0	1
1	0	1	1
0	1	1	1

Table 5.18 Final test tube of CLIQUES1.

v_1	v_2	v_3	v_4
1	1	1	0

k-sets form k-cliques. For this, it delivers all subgraphs of G induced by the subsets of $\binom{k}{2}$ edges. Among these subgraphs, those subgraphs are determined that contain k vertices. These are exactly the k-cliques. The algorithm requires $4m + 2n(k + 1) - k^2 - k$ steps.

Algorithm 5.22 CLIQUES2(T, m, n, k)

Input: $[m + n, \binom{\binom{m}{k}}{2}]$ library T

1: $T \leftarrow$ EdgeInducedGraphs (T, m, n)
2: $T \leftarrow$ Weightening (T, m, n, k)
3: **if** ¬empty (T) **then**
4:     **return** T
5: **else**
6:     report "no k-clique"
7: **end if**

Example 5.18. Consider the graph G in Figure 5.8 and the initial library T with $k = 3$ giving rise to the memory complexes in Table 5.19. EDGEINDUCEDGRAPHS yields the test tube in Table 5.20, and WEIGHTENING provides the memory strands shown in the first row. Thus, the graph contains the single 3-clique $\{v_1, v_2, v_3\}$.   ◇

Independent Sets

The algorithm CLIQUES1 can be modified to yield the independent k-sets in a given graph G. The input of the algorithm is an $[n, \binom{n}{k}]$ library T, providing encoded DNA of all k-subsets of vertices, $1 \leq k \leq n$. In the algorithm CLIQUES1, line 6 is replaced as follows,

Table 5.19 Initial test tube of CLIQUES2.

e_1	e_2	e_3	e_4	v_1	v_2	v_3	v_4
1	1	1	0	0	0	0	0
1	1	0	1	0	0	0	0
1	0	1	1	0	0	0	0
0	1	1	1	0	0	0	0

Table 5.20 Test tube produced by EDGEINDUCEDGRAPHS in CLIQUES2.

e_1	e_2	e_3	e_4	v_1	v_2	v_3	v_4
1	1	1	0	1	1	1	0
1	1	0	1	1	1	1	1
1	0	1	1	1	1	1	1
0	1	1	1	1	1	1	1

6:    **if** not (adjacent (v_i, v_j)) **then** .

Thus, the tube T^{++} yields all memory complexes in which the ith and jth vertex are not adjacent. If not, the memory complexes are filtered out. This algorithm requires $n^2 + 5n$ steps. The maximum independent set problem can be solved by invoking the algorithm for decreasing parameters $k = n, n-1, \ldots$ until the first non-empty output test tube is encountered.

Example 5.19. Take the graph G in Figure 5.8 and the library T with $k = 2$. The final test tube illustrated in Table. 5.21 delivers the independent 2-subsets $\{v_1, v_4\}$ and $\{v_2, v_4\}$ of G. $\diamond$

Here is another algorithm for the problem of independent sets. For this, the input of the algorithm is an $[n + m, \binom{n}{k}]$ library T, providing encoded DNA of all k-subsets of vertices in a graph G, where $1 \leq k \leq n$. The algorithm determines which of the k-sets are independent. To this end, it computes the subgraphs in G induced by the k-subsets of edges. This is implemented by VERTEXINDUCEDGRAPHS. Among these subgraphs, those subgraphs are determined that contain no edges. This is implemented by WEIGHTENING. The algorithm requires $4m + 2n$ steps.

Algorithm 5.23 INDEPENDENTSETS(T, m, n, k)

Input: $[m + n, \binom{n}{k}]$ library T
1: $T \leftarrow$ VertexInducedGraphs (T, n, m)
2: $T \leftarrow$ Weightening $(T, n, m, 0)$
3: **if** $\neg$empty (T) **then**
4:     **return** T
5: **else**
6:     report "no k-independent set"
7: **end if**

Colorings

The algorithm 3-COLORINGS solves the 3-coloring problem in a graph G. The input of the algorithm is an $[n + 2n, \sum\limits_{i=1}^{\lfloor n/3 \rfloor} \binom{n}{i}]$ library T, providing encoded

Table 5.21 Final test tube of INDEPENDENTSETS.

v_1	v_2	v_3	v_4
1	0	0	1
0	1	0	1

DNA of all $\lfloor n/3 \rfloor$-subsets of vertices. The algorithm decomposes the vertex set of G into three disjoint independent subsets so that each subset can be given a different color. First, INDEPENDENTSETS filters out those $\lfloor n/3 \rfloor$-subsets of vertices that are independent. Second, COMPLEMENT yields the complement of each independent subset storing the complementary set in the second n substrands. Third, INDEPENDENTSUBSETS constructs an independent subset of the subgraphs spanned by those vertices that are given by the second n substrands. Fourth, INDEPENDENTSETS eliminates those memory complexes in which vertices given by the third n substrands are adjacent. The algorithm requires $3n^2 + 15n$ steps.

Algorithm 5.24 3-COLORINGS(T, n)

Input: $[n + 2n, \sum_{i=1}^{\lfloor n/3 \rfloor} \binom{n}{i}]$ library T

1: $T \leftarrow$ IndependentSets $(T, 1, n)$
2: $T \leftarrow$ Complement (T, n)
3: $T \leftarrow$ IndependentSubsets $(T, n + 1, 2n, n)$
4: $T \leftarrow$ IndependentSets $(T, 2n + 1, 3n)$
5: **if** ¬empty (T) **then**
6:     **return** T
7: **else**
8:     report "no 3-colored subgraph"
9: **end if**

Example 5.20. Take the graph G in Figure 5.8 and consider the initial test tube T in Table 5.22. INDEPENDETSETS leaves the tube invariant and COMPLEMENT yields the test tube in Table 5.23. After this, INDEPENDENTSUBSETS delivers independent subsets as described in Table 5.24.

Table 5.22 Initial test tube of 3-COLORINGS.

v_1	v_2	v_3	v_4	v_1	v_2	v_3	v_4	v_1	v_2	v_3	v_4
1	0	0	0	0	0	0	0	0	0	0	0
0	1	0	0	0	0	0	0	0	0	0	0
0	0	1	0	0	0	0	0	0	0	0	0
0	0	0	1	0	0	0	0	0	0	0	0

Table 5.23 Test tube after INDEPENDENTSETS and COMPLEMENT in 3-COLORINGS.

v_1	v_2	v_3	v_4	v_1	v_2	v_3	v_4	v_1	v_2	v_3	v_4
1	0	0	0	0	1	1	1	0	0	0	0
0	1	0	0	1	0	1	1	0	0	0	0
0	0	1	0	1	1	0	1	0	0	0	0
0	0	0	1	1	1	1	0	0	0	0	0

Table 5.24 Test tube after INDEPENDENTSUBSETS in 3-COLORINGS.

v_1	v_2	v_3	v_4	v_1	v_2	v_3	v_4	v_1	v_2	v_3	v_4
1	0	0	0	0	1	0	1	0	0	1	0
0	1	0	0	1	0	0	1	0	0	1	0
0	0	1	0	1	0	0	1	0	1	0	0
0	0	0	1	1	0	0	0	0	1	1	0

Table 5.25 Final test tube of 3-COLORINGS.

v_1	v_2	v_3	v_4	v_1	v_2	v_3	v_4	v_1	v_2	v_3	v_4
1	0	0	0	0	1	0	1	0	0	1	0
0	1	0	0	1	0	0	1	0	0	1	0
0	0	1	0	1	0	0	1	0	1	0	0

Finally, INDEPENDETSETS yields the final test tube illustrated in Table 5.25. For instance, the first memory complex in the final test tube encodes a decomposition of the vertex set into three independent subsets: $\{v_1\}$, $\{v_2, v_4\}$, and $\{v_3\}$. ◇

Bipartite Subgraphs

The algorithm BIPARTITESUBGRAPHS determines whether a graph G has a k-bipartite subgraph. The input of the algorithm is a $[2n, \binom{2n}{k}]$ library T. Each memory complex in T encodes two subsets of the vertices in G, the first and second of which are given by the first and second n substrands, respectively. Thus, each memory complex corresponds to a union of a k-subset of the vertex set into two disjoint sets. First, both subsets of such a k-subset are subject to INDEPENDENTSETS, filtering out those subsets that are independent in G. Second, memory complexes in which both subsets are independent in G serve as input of MUTUALLYEXCLUSIVESETS, delivering those memory complexes in which the independent subsets are disjoint. Hence, the final memory complexes encode k-bipartite subgraphs of G. The algorithm requires $2n^2 + 10n + 6n = 2n^2 + 16n$ steps.

Algorithm 5.25 BIPARTITESUBGRAPHS(T, n, k)

Input: $[2n, \binom{2n}{k}]$ library T
1: $T \leftarrow$ IndependentSets $(T, 1, n)$
2: $T \leftarrow$ IndependentSets $(T, n + 1, 2n)$
3: $T \leftarrow$ MutuallyDisjointSets (T, n)
4: **if** ¬empty (T) **then**
5:     **return** T
6: **else**
7:     report "no k-bipartite subgraph"
8: **end if**

Fig. 5.10 A bipartite graph G.

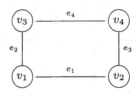

Table 5.26 Initial test tube of BIPARTITESUBGRAPHS.

v_1	v_2	v_3	v_4	v_1	v_2	v_3	v_4
1	1	1	1	0	0	0	0
1	1	0	0	1	1	0	0
0	0	1	1	0	0	1	1
1	0	1	0	1	0	1	0
0	1	0	1	0	1	0	1
1	0	0	1	0	1	1	0
0	1	1	0	1	0	0	1
0	0	0	0	1	1	1	1
1	1	0	0	0	0	1	1
0	0	1	1	1	1	0	0

Example 5.21. Consider a test tube T with $n = 4$ and $k = 4$ holding $\binom{8}{4}$ different types of memory complexes. To simplify the example, assume that the test tube T consists of the memory complexes in Table 5.26. In terms of the graph G in Figure 5.10, the first INDEPENDENTSETS statement yields the tube in Table 5.27, while the second INDEPENDENTSETS statement provides the tube in Table 5.28. Then MUTUALLYDISJOINTSETS gives the final tube illustrated in Table 5.29. Thus the graph G is a 4-bipartite graph with vertex subsets $\{v_1, v_4\}$ and $\{v_2, v_3\}$.   ◇

Table 5.27 Test tube after first INDEPENDENTSETS in BIPARTITESUBGRAPHS.

v_1	v_2	v_3	v_4	v_1	v_2	v_3	v_4
1	0	0	1	0	1	1	0
0	1	1	0	1	0	0	1
0	0	0	0	1	1	1	1

Table 5.28 Test tube after second INDEPENDENTSETS in BIPARTITESUBGRAPHS.

v_1	v_2	v_3	v_4	v_1	v_2	v_3	v_4
1	0	0	1	0	1	1	0
0	1	1	0	1	0	0	1

Table 5.29 Final test tube of BIPARTITE SUBGRAPHS.

v_1	v_2	v_3	v_4	v_1	v_2	v_3	v_4
1	0	0	1	0	1	1	0
0	1	1	0	1	0	0	1

Matchings

The algorithm MATCHINGS solves the k-matching problem in a graph G. The input of the algorithm is an $[m+n, \binom{m}{k}]$ library T, providing encoded DNA of all k-subsets of edges, where $1 \leq k \leq n/2$. First, the induced subgraphs from all k-subsets of edges are calculated. Such a subgraph yields a k-matching if and only if it has $2k$ vertices. Thus, WEIGHTENING filters out those subgraphs with $2k$ vertices. The algorithm requires $4m + 2n(2k + 1) - 4k^2 - 2k$ steps.

Algorithm 5.26 MATCHINGS(T, m, n, k)

Input: $[m + n, \binom{m}{k}]$ library T
1: $T \leftarrow$ EdgeInducedGraphs (T, m)
2: $T \leftarrow$ Weightening $(T, m, n, 2k)$
3: **if** ¬empty (T) **then**
4:     **return** T
5: **else**
6:     report "no k-matching"
7: **end if**

Example 5.22. Take the graph G in Figure 5.8 and consider the initial test tube T with $k = 2$ given in Table 5.30. EDGEINDUCEDGRAPHS yields the memory complexes shown in Table 5.31, and WEIGHTENING provides the final test tube illustrated in Table 5.32. Thus, the subgraph induced by the edges e_1 and e_4 is the only 2-matching in G.   ◇

Table 5.30 Initial test tube of MATCHINGS.

e_1	e_2	e_3	e_4	v_1	v_2	v_3	v_4
1	1	0	0	0	0	0	0
1	0	1	0	0	0	0	0
1	0	0	1	0	0	0	0
0	1	1	0	0	0	0	0
0	1	0	1	0	0	0	0
0	0	1	1	0	0	0	0

Table 5.31 Test tube after EDGEINDUCEDGRAPHS in MATCHINGS.

e_1	e_2	e_3	e_4	v_1	v_2	v_3	v_4
1	1	0	0	1	1	1	0
1	0	1	0	1	1	1	0
1	0	0	1	1	1	1	1
0	1	1	0	1	1	1	0
0	1	0	1	1	0	1	1
0	0	1	1	0	1	1	1

Table 5.32 Final test tube of MATCHINGS.

e_1	e_2	e_3	e_4	v_1	v_2	v_3	v_4
1	0	0	1	1	1	1	1

Perfect Matchings

A small change in the MATCHINGS algorithm solves the perfect matching
problem. Let $n > 0$ be an even integer. The input of the algorithm is an
$[m + n, \binom{m}{n/2}]$ library T, providing encoded DNA of all $n/2$-subsets of edges.
In opposition to the MATCHINGS algorithm, WEIGHTENING receives n as the
last parameter instead of $2k$. The algorithm requires $4m + 2n(n+1) - n^2 - n =
4m + n^2 + n$ steps.

Algorithm 5.27 PERFECTMATCHINGS(T, m, n)

Input: $[m + n, \binom{m}{n/2}]$ library T
1: $T \leftarrow$ EdgeInducedGraphs (T, m)
2: $T \leftarrow$ Weightening (T, m, n, n)
3: **if** ¬empty (T) **then**
4:     **return** T
5: **else**
6:    report "no perfect matching"
7: **end if**

Edge-Dominating Sets

The algorithm EDGEDOMINATINGSETS calculates all dominating edge sets
of size k in a graph G. The input of the algorithm is an $[m + (n + m), \binom{m}{k}]$
library T, providing encoded DNA of all k-subsets of edges, where $1 \leq k \leq m$.
First, EDGEINDUCEDGRAPHS provides the subgraphs in G, which are induced
by the k-subsets of edges. For this, the vertex set of such a subgraph is

stored in the second set of substrands of size n. Second, INCIDENCERELATION calculates the set of incident edges for the vertex set of such a subgraph. These edges are stored in the third set of substrands of size m. Such a set of incident edges is k-dominating in G if and only if it equals the edge set in G. Therefore, WEIGHTENING filters out those memory complexes whose m substrands on the third set are turned on. The algorithm requires $4m + nm + 2n + 2m(m+1) - m^2 - m = n(m+2) + m^2 + m$ steps. The minimum edge dominating k-set problem can be solved by invoking this algorithm for increasing parameters $k = 1, \ldots, n$ (respectively input test tubes) until the corresponding output test tube is non-empty.

Algorithm 5.28 EDGEDOMINATINGSETS(T, m, n, k)

Input: $[m + (n + m), \binom{m}{k}]$ library T
1: $T \leftarrow$ EdgeInducedGraphs (T, m)
2: $T \leftarrow$ IncidenceRelation $(T, m + 1, m + n, m, n)$
3: $T \leftarrow$ Weightening $(T, m + n, m, m)$
4: **if** ¬empty (T) **then**
5:      **return** T
6: **else**
7:      report "no k-edge-dominating set"
8: **end if**

Example 5.23. Start with the initial test tube T with $n = 4$, $m = 4$, and $k = 1$ described in Table 5.33. In terms of the graph G in Figure 5.8, EDGEINDUCEDGRAPHS produces the tube in Table 5.34. Then INCIDENCERELATION yields the tube in Table 5.35. Finally, WEIGHTENING gives the final test tube in which the last m substrands are turned on as shown in Table 5.35. Thus, $\{e_2\}$ and $\{e_3\}$ are 1-dominating edge sets in G. $\diamond$

Table 5.33 Initial test tube of EDGEDOMINATINGSETS.

e_1	e_2	e_3	e_4	v_1	v_2	v_3	v_4	e_1	e_2	e_3	e_4
1	0	0	0	0	0	0	0	0	0	0	0
0	1	0	0	0	0	0	0	0	0	0	0
0	0	1	0	0	0	0	0	0	0	0	0
0	0	0	1	0	0	0	0	0	0	0	0

Table 5.34 Test tube after EDGEINDUCEDGRAPHS in EDGEDOMINATINGSETS.

e_1	e_2	e_3	e_4	v_1	v_2	v_3	v_4	e_1	e_2	e_3	e_4
1	0	0	0	1	1	0	0	0	0	0	0
0	1	0	0	1	0	1	0	0	0	0	0
0	0	1	0	0	1	1	0	0	0	0	0
0	0	0	1	0	0	1	1	0	0	0	0

Table 5.35 Test tube after INCIDENCERELATION in EDGEDOMINATINGSETS.

e_1	e_2	e_3	e_4	v_1	v_2	v_3	v_4	e_1	e_2	e_3	e_4
1	0	0	0	1	1	0	0	1	1	1	0
0	1	0	0	1	0	1	0	1	1	1	1
0	0	1	0	0	1	1	0	1	1	1	1
0	0	0	1	0	0	1	1	0	1	1	1

Table 5.36 Final test tube of EDGEDOMINATINGSETS.

e_1	e_2	e_3	e_4	v_1	v_2	v_3	v_4	e_1	e_2	e_3	e_4
0	1	0	0	1	0	1	0	1	1	1	1
0	0	1	0	0	1	1	0	1	1	1	1

Hamiltonian Paths

The sticker algorithm HAMILTONIANPATHS determines all Hamiltonian paths in a graph. The input of the algorithm is an $[m + 2n, \binom{m}{n-1}]$ library T, providing encoded DNA of all $n - 1$-subsets of the edge set, where $n \leq m + 1$. The algorithm determines which of these sets form Hamiltonian paths. To do this, it proceeds in the following steps:

- The induced subgraphs from the $n - 1$-sets of edges are calculated by EDGEINDUCEDGRAPHS.
- The spanning trees of the graph are computed by SPANNINGTREES.
- For each spanning tree, the degree modulo 2 of each vertex is computed (statements 3-11).
- The spanning trees whose (unordered) degree sequence taken modulo 2 contains two entries 1 and $n-2$ entries 0 are determined by WEIGHTENING.
- The return test tube of WEIGHTENING is the output of the algorithm.

The algorithm requires $4m + 4m + n + 8m + 6n - 6 = 16m + 6n - 6$ steps.

The algorithm yields all Hamiltonian paths in a graph G. To see this, observe that statement 2 yields all spanning trees of the graph.

Claim that a spanning tree is a Hamiltonian path if and only if its degree sequence taken modulo 2 has two ones and $n - 2$ zeros (i.e., the sequence is of the form $0\ldots010\ldots010\ldots0$). Indeed, it is clear that each Hamiltonian path has such a degree sequence taken modulo 2. Conversely, consider a spanning tree T whose degree sequence taken modulo 2 is of the form $0\ldots010\ldots010\ldots0$. As each tree has at least two vertices of degree one, it follows that the two ones in the degree sequence stem from the vertices of degree one. All other vertices of T have even degree (greater than zero as the graph is connected). But by Theorem 2.18, a tree with n vertices has $n - 1$ edges and thus in view of the Handshaking Lemma, the sum of

---

Algorithm 5.29 HAMILTONIANPATHS(T, m, n)

Input: $[m + 2n, \binom{m}{n-1}]$ library T

1:   $T \leftarrow$ EdgeInducedGraphs (T, m, n)
2:   $T \leftarrow$ SpanningTrees (T, m, n)
3:   **for** $i \leftarrow 1$ to m **do**
4:     separate (T, T^+, T^-, i)
5:     **for** $j \leftarrow 1$ to 2 **do**
6:       separate $(T^+, T^{++}, T^{+-}, m + n + i_j)$
7:       set $(T^{+-}, m + n + i_j)$
8:       clear $(T^{++}, m + n + i_j)$
9:     **end for**
10:   merge (T^-, T^{++}, T^{+-}, T)
11: **end for**
12: $T \leftarrow$ Weightening $(T, m + n, n, 2)$
13: **if** ¬empty (T) **then**
14:    **return** T
15: **else**
16:    report "no Hamiltonian path"
17: **end if**

---

degrees of all vertices in T equals $2(n-1)$. Thus, each of the $n-2$ vertices of even degree must have degree 2. Hence, T has (unordered) degree sequence $2 \ldots 212 \ldots 212 \ldots 2$.

Finally, assume that we are travelling along the edges of the spanning tree T as a travelling salesman would do. For this, we start at one of the two vertices of degree one, say v_{i_1}, and move in T to the unique adjacent vertex, say v_{i_2}. This vertex may also have degree one. But if $n > 2$, then the spanning tree would decompose into two components, the first of which is given by the edge $v_{i_1} v_{i_2}$ and the second of which is defined by a simple cycle of the remaining vertices. In view of the degrees, these subgraphs would have no vertex in common, contradicting the hypothesis that the subgraph T is connected. By proceeding in this way, we see that the spanning tree T forms a Hamiltonian path. This proves the claim, and the correctness of the algorithm follows.

Example 5.24. Given the graph in Figure 5.8 and the corresponding spanning trees in Figure 5.9, HAMILTONIANPATHS yields the memory complexes in Table 5.37, where the last four substrands indicate the degrees modulo 2 of the vertices. It follows that the first two memory complexes correspond to Hamiltonian paths. $\diamond$

Hamiltonian Cycles

The sticker algorithm HAMILTONIANCYCLES determines all Hamiltonian cycles in a graph. The Hamiltonian cycles are formed exactly by those Hamiltonian paths, whose ending vertices are adjacent. Therefore, we may first

Table 5.37 Final test tube of HAMILTONIANPATHS.

e_1	e_2	e_3	e_4	v_1	v_2	v_3	v_4	$d(v_1)_2$	$d(v_2)_2$	$d(v_3)_2$	$d(v_4)_2$
1	1	0	1	1	1	1	1	0	1	0	1
1	0	1	1	1	1	1	1	1	0	0	1
0	1	1	1	1	1	1	1	1	1	1	1

employ the algorithm HAMILTONIANPATHS to calculate all Hamiltonian paths in the graph. The Hamiltonian paths correspond to those memory complexes of length $m + 2n$ whose last n substrands contain two entries 1 and $n - 2$ entries 0. The two non-zero entries specify the vertices of degree 1 in the Hamiltonian path. If these two non-zero entries are given by the substrands $m + n + i_1$ and $m + n + i_2$, with $1 \leq i_1, i_2 \leq n$, then a corresponding Hamiltonian cycle exists if and only if the graph contains the edge $e_i = v_{i_1} v_{i_2}$. For each edge in the graph, all Hamiltonian paths are checked in parallel to determine whether this condition is fulfilled. The initially empty test tube T_1 contains the Hamiltonian cycles determined so far. The algorithm requires $16m + 6n - 6 + 4m = 20m + 6n - 6$ steps.

Algorithm 5.30 HAMILTONIANCYCLES(T, m, n)

Input: $[m + 2n, \binom{m}{n-1})]$ library T, test tube T_1 initially empty
1: $T \leftarrow$ HamiltonianPaths (T, m, n)
2: **for** $i \leftarrow 1$ to m **do**
3:     separate $(T, T^+, T^-, m + n + i_1)$
4:     separate $(T^+, T^{++}, T^{+-}, m + n + i_2)$
5:     merge (T^{++}, T_1)
6:     merge (T^-, T^{+-}, T)
7: **end for**
8: **if** $\neg$empty (T_1) **then**
9:     **return** T_1
10: **else**
11:     report "no Hamiltonian cycle"
12: **end if**

For instance, the graph in Figure 5.8 has no Hamiltonian cycle since none of its Hamiltonian paths can be closed to a cycle.

Steiner Trees

Finally, sticker algorithms for two Steiner tree problems are provided. For this, let Z be a subset of the vertex set of a connected graph G. The first problem is to determine all Steiner trees $H = (U, F)$ for Z in G so that F

is a k-subset of the edge set. For this, we relabel the vertices in G so that $Z = \{v_1, \ldots, v_\ell\}$. The input of the algorithm is an $[m + n, \binom{m}{k}]$ library T, providing encoded DNA of all k-subsets of edges in G, where $1 \leq k \leq m$. The algorithm proceeds in the following steps:

- The induced subgraphs from the k-subsets of edges are calculated.
- Those induced subgraphs are determined that have $k + 1$ vertices (statements 2–5). These subgraphs are minimally connected and thus form subtrees in G.
- Those subgraphs are extracted that contain all vertices in Z (statements 6–9).
- The test tube T provides the output of the algorithm.

The algorithm requires $4m + 2n + 2\ell$ steps.

Algorithm 5.31 STEINERTREES(T, m, n, ℓ, k)

Input: $[m + n, \binom{m}{k}]$ library T – first problem
1:   $T \leftarrow$ EdgeInducedGraphs (T, m, n)
2:   **for** $i \leftarrow 1$ to n **do**
3:      separate $(T, T^+, T^-, m + i)$
4:      $T \leftarrow T^+$
5:   **end for**
6:   **for** $i \leftarrow 1$ to ℓ **do**
7:      separate $(T, T^+, T^-, m + i)$
8:      $T \leftarrow T^+$
9:   **end for**
10: **if** ¬empty (T) **then**
11:     **return** T
12: **else**
13:     report "no k-Steiner tree"
14: **end if**

The second problem is to determine all Steiner trees $H = (U, F)$ for Z in G so that $U \backslash Z$ is a k-set. This problem can be solved in the same way as the first problem. The input of the algorithm is an $[m+n, \binom{m}{k+\ell-1}]$ library T, providing encoded DNA of all $k + \ell - 1$-subsets of edges in G, where $1 \leq k + l - 1 \leq m$. Those induced subgraphs that have $k + \ell$ vertices are minimally connected and thus form subtrees in G. Those subtrees that contain all vertices in Z provide the solution. Therefore, the algorithm STEINERTREES applied to the above input test tube solves the second problem.

Limitation of DNA Algorithms

There are more errors than just the stickers binding in the wrong places. DNA strands are flexible and will fold back on themselves, thereby forming secondary structures. This will make regions in a memory strand inaccessible for proper use. This problem can be avoided by composing memory strands of only pyrimidines and sticker strands only of purines, or vice versa.

DNA algorithms in general are limited by several key factors. A first limiting factor is the length of the memory complexes: oligonucleotides longer than 15,000 nt might be fragmented by the shear forces of pouring and mixing the test tubes. A reasonable problem size for sticker algrithms is $l = 12,000$ nt for memory strands and $m = 20$ nt for sticker strands. This will allow the representation of binary numbers of length $n = 600$ bits.

A second limiting factor is the speed of the DNA operations. For instance, separated complementary strands of purified DNA recognize each other as a result of collisions. Under appropriate conditions, they specifically reassociate or anneal to a double-stranded molecule. Reassociation can be measured by the C_0t parameter (pronouced "cot") introduced by R. Britten et al. in the 1960s. The DNA of each organism may be characterized by the value of C_0t at which the reassociation reaction is half completed under controlled conditions. A C_0t of 1 mole of nucleotides times seconds per liter results if DNA is incubated for 1 hour at a concentration of 83 µg/ml. The C_0t values range over several orders of magnitude depending on the type (genome) of the DNA.

A third limiting factor comes from the fact that DNA operations are error prone. The most erroneous sticker operation is separate. Consider a sticker algorithm with S separate operations, with the other operations being negligible. Assume that each separate operation takes one unit of time to complete, no strands are physically lost during the separation process, and there is a common probability p of correctly processing each complex, which is assumed to be near unity. The computation will then take S units of time and the fraction of complexes not correctly processed will be depressingly high $\delta = 1 - p^S$. For instance, if $p = 0.9$ and $S = 100$ then $\delta = 0.99997$.

There are techniques to make the fraction δ smaller using intelligent space and time trade-offs. This requires no changes of the quantities p and S. Each separate operation can be repeated M times. This time slowdown improves the error fraction δ from $1 - p^S$ to $(1 - p^S)^M$. In addition, each separate operation can be performed on $H = 2N + 1$ test tubes T_i, $-N \leq i \leq N$. This *compound separate* operation, implemented by the algorithm COMPOUNDSEPARATE, requires that the test tubes are organized so that if no error is committed, the memory complexes with ith bit on located in test tube T_j move into test tube T_{j+1}, while those with ith bit off move into test tube T_{j-1}, with absorption at the boundary tubes. In this way, the memory complexes perform a biased random walk in tubes T_{-N} through T_N so that most memory complexes with ith bit on will end up in T_N, while most memory complexes with ith bit off will end up in T_{-N}. This process resembles the gambler's ruin problem studied in quantitative finance.

Algorithm 5.32 COMPOUNDSEPARATE(T, N, Q, k)

Input: tube $T = T_0$, extra tubes $T_{-N}, \ldots, T_{-1}$ and $T_1, \ldots, T_N$ initially empty,
    integers $k \geq 0$ and $Q \geq 1$
1: **for** $i \leftarrow 1$ to Q **do**
2:     **for** $j \leftarrow -N + 1$ to $N - 1$ **do**
3:       **if** $i + j \equiv 1 \bmod 2$ **then**
4:         separate (T_j, T^+, T^-, k)
5:         merge (T^+, T_{j+1})
6:         merge (T^-, T_{j-1})
7:       **end if**
8:     **end for**
9: **end for**

Theorem 5.25. *Consider a gambler who wins or loses a dollar with probabilities p and q, respectively. Let his initial capital be $z > 0$ and let him play against an adversary with initial capital $a - z > 0$ so that the combined capital is a. The game continues until the gambler's capital either is reduced to zero or has increased to a. The probability of the gambler's ultimate ruin is given by*

$$q_z = \begin{cases} \frac{(q/p)^a - (q/p)^z}{(q/p)^a - 1} & \text{if } p \neq q, \\ 1 - \frac{z}{a} & \text{otherwise.} \end{cases} \tag{5.5}$$

The probability p_z of winning satisfies $p_z + q_z = 1$, and the solution (q_z, p_z) is uniquely determined.

A game against an infinitely rich adversary $(a \to \infty)$ leads to

$$q_z = \begin{cases} 1 & \text{if } p \leq q, \\ (q/p)^z & \text{otherwise.} \end{cases} \tag{5.6}$$

The expected duration of the game is given by

$$D_z = \begin{cases} \frac{z}{q-p} - \frac{a}{q-p} \cdot \frac{1 - (q/p)^z}{1 - (q/p)^a} & \text{if } p \neq q, \\ z(a - z) & \text{otherwise.} \end{cases} \tag{5.7}$$

Proof. After the first trial the gambler's fortune is either $z - 1$ or $z + 1$. Therefore, the difference equation

$$q_z = p q_{z+1} + q q_{z-1}, \quad 1 < z < a, \tag{5.8}$$

with boundary conditions $q_0 = 1$ and $q_a = 0$ holds. Let $p \neq q$. The difference equation (5.8) has the solutions $q_z = 1$ and $q_z = (q/p)^z$. The linear combination of both solutions with arbitrary constants A and B also provides a solution

$$q_z = A + B \left(\frac{q}{p}\right)^z. \tag{5.9}$$

The boundary conditions imply that $A + B = 1$ and $A + B(q/p)^a = 0$, and hence the term (5.5), forms a solution of the difference equation (5.8) satisfying the boundary conditions.

If $p = q = 1/2$, then $q_z = 1$ and $q_z = z$ are solutions of the difference equation (5.8), and therefore $q_z = A + Bz$ is a solution, where A and B are arbitrary constants. The boundary conditions require that $A = 1$ and $A + Ba = 0$, and hence $q_z = 1 - z/a$ is a solution of the difference equation (5.8) satisfying the boundary conditions.

The probability p_z of the gambler's winning equals the probability of the adversary's ruin and is therefore obtained from the above formulas on replacing p, q and z by q, p and $a - z$, respectively. From this it can be readily seen that $p_z + q_z = 1$ holds. Claim that the solution is unique. Indeed, given an arbitrary solution of Eq. (5.8), the constants A and B can be chosen so that Eq. (5.9) will agree with it for $z = 0$ and $z = 1$. From these two values all other values can be found by substituting in Eq. (5.8) successively $z = 1, 2, 3, \ldots$.

The limiting case $a = \infty$ immediately follows from Eq. (5.5).

Finally, if the gambler's first trial results in success the expected duration is $D_{z+1} + 1$. Therefore, the expected duration D_z satisfies the non-homogeneous difference equation

$$D_z = pD_{z+1} + qD_{z-1} + 1, \quad 0 < z < a, \tag{5.10}$$

with boundary conditions $D_0 = 0$ and $D_a = 0$. If $p \neq q$, then $D_z = z/(q-p)$ is a solution of Eq. (5.10). The difference Δ_z of any two solutions satisfies the homogeneous equation $\Delta_z = p\Delta_{z+1} + q\Delta_{z-1}$, and we already know that all solutions of this equation are of the form $A + B(q/p)^z$. It follows that all solutions of Eq. (5.10) have the shape

$$D_z = \frac{z}{q-p} + A + B\left(\frac{q}{p}\right)^z. \tag{5.11}$$

The boundary conditions require that $A + B = 0$ and $A + B(q/p)^a = -a/(q-p)$. Solving for A and B, we find the first equation in Eq. (5.7).

If $p = q = 1/2$, then $q_z = -z^2$ is a solution of Eq. (5.10), and all solutions of Eq. (5.10) are of the form $D_z = -z^2 + A + Bz$. The solution D_z satisfying the boundary conditions is $D_z = z(a - z)$, as needed.   □

Let p be the probability that a separation step correctly moves a complex. If each complex continues to be processed until it reaches either T_{-N} or T_N, then by setting $a = 2N$ and $z = N$, Eq. (5.6) yields the probability of correctly processing a complex,

$$p_\infty = 1 - \left(\frac{1-p}{p}\right)^N. \tag{5.12}$$

Moreover, Eq. (5.7) provides the expected time for a complex to arrive in either T_{-N} or T_N. If $p \neq \frac{1}{2}$ then

$$\tau_N = \frac{N}{1 - 2p} \left(1 - 2 \frac{1 - r^N}{1 - r^{2N}} \right), \quad r = \frac{1 - p}{p}, \tag{5.13}$$

and if $p = \frac{1}{2}$, then $\tau_N = N^2$. For instance, if $p = 0.9$ and $N = 5$, then $p_\infty = 0.99998$ and the expected time is $\tau_5 = 6.25$.

5.4 Splicing Systems

Splicing systems are generative mechanisms based on the splicing operation formalized by T. Head (1987) as a model of DNA recombination. It will be proved that the generative power of splicing systems equals that of Turing machines and that universal splicing systems exist that are able to simulate any Turing program.

5.4.1 Basic Splicing Systems

The splicing system model is based on an operation specific to DNA recombination, which can be thought of as a combination of cutting double-stranded DNA molecules and reconnecting the cut parts to obtain new DNA molecules. This operation, termed *splicing*, should not be confused with the splicing operation used in gene transcription in eukaryotic cells.

Example 5.26. Consider two double-stranded DNA molecules

$$5' - \text{CCCCCTCGACCCCC} - 3' \quad 5' - \text{AAAAATCGAAAAAA} - 3'$$
$$3' - \text{GGGGGAGCTGGGGG} - 5' \quad 3' - \text{TTTTTAGCTTTTTT} - 5'$$

and the restriction enzyme TaqI with the recognition site

$$5' - \text{TCGA} - 3'$$
$$3' - \text{AGCT} - 5' \, .$$

This enzyme cuts the DNA molecules at its recognition sites

$$5' - \text{CCCCCT} \quad \text{CGACCCCC} - 3' \quad 5' - \text{AAAAAT} \quad \text{CGAAAAAA} - 3'$$
$$3' - \text{GGGGGAGC} \quad \text{TGGGGG} - 5' \quad 3' - \text{TTTTTAGC} \quad \text{TTTTTT} - 5' \, .$$

Recombination or splicing of molecules one and four and molecules two and three yields the double-stranded molecules

$$5' - \text{CCCCCTCGAAAAAA} - 3' \quad 5' - \text{AAAAATCGACCCCC} - 3'$$
$$3' - \text{GGGGGAGCTTTTTT} - 5' \quad 3' - \text{TTTTTAGCTGGGGG} - 5' \, .$$

$\diamond$

A *splicing system* is a quadruple $S = (\Sigma, V, R, A)$ consisting of a finite set of terminal symbols Σ, a finite set of non-terminal symbols V with $\Sigma \cap V = \emptyset$, a finite set of *splicing rules* $R \subseteq \Gamma^* \# \Gamma^* \$ \Gamma^* \# \Gamma^*$, where $\Gamma = \Sigma \cup V$ and $\#$ and $\$$ are special symbols not contained in Γ, and a finite set of axioms $A \subseteq \Gamma^*$.

The splicing rules are used to provide one-step derivations. For this, $r = w_1 \# w_2 \$ w_3 \# w_4 \in R$ and define $(u, v) \vdash_r (x, y)$, $u, v, x, y \in \Gamma^*$, if there are strings $u_1, u_2, v_1, v_2 \in \Gamma^*$ so that

$$
\begin{aligned}
u &= u_1 w_1 w_2 u_2, \\
v &= v_1 w_3 w_4 v_2, \\
x &= u_1 w_1 w_4 v_2, \\
y &= v_1 w_3 w_2 u_2 \, .
\end{aligned}
\tag{5.14}
$$

Example 5.27. Consider the splicing system $S = (\Sigma, V, R, A)$ with $\Sigma = \{a, b, c\}$, $V = \emptyset$, $R = \{r_b = b \# \$ b \#, r_c = c \# \$ c \#\}$ and $A = \{abaca, acaba\}$. We have

$$(ab|aca, acab|a) \vdash_{r_b} (aba, acabaca), \tag{5.15}$$
$$(abac|a, ac|aba) \vdash_{r_c} (abacaba, aca), \tag{5.16}$$

where vertical bars indicate the substrings. $\diamond$

The splicing rules provide a basic tool for building a generative mechanism. To this end, the language generated consists of the strings over Σ, which are iteratively obtained by applying the rules to the axioms and the strings obtained in preceding splicing steps. More precisely, for any language $L \subseteq \Gamma^*$, define

$$
\sigma(L) = \\
\{x \in \Gamma^* \mid (u, v) \vdash_r (x, y) \text{ or } (u, v) \vdash_r (y, x), \; u, v \in L, \; r \in R\} \, . \tag{5.17}
$$

Moreover, let

$$\sigma^*(L) = \bigcup_{n \geq 0} \sigma^n(L), \tag{5.18}$$

where

$$\sigma^0(L) = L, \tag{5.19}$$
$$\sigma^{n+1}(L) = \sigma^n(L) \cup \sigma(\sigma^n(L)), \quad n \geq 0. \tag{5.20}$$

The *language* of the splicing system S is defined as

$$L(S) = \sigma^*(A) \cap \Sigma^* .\tag{5.21}$$

Example 5.28. Reconsider the splicing system S in Example 5.27. By using the strings on the right hand side in Eqs. (5.15) and (5.16), we obtain

$$(abacab|a, ab|aca) \vdash_{r_b} (abacabaca, aba),\tag{5.22}$$
$$(acabac|a, ac|aba) \vdash_{r_c} (acabacaba, aca) .\tag{5.23}$$

More generally, for each $n \geq 1$,

$$((abac)^n ab|a, ab|aca) \vdash_{r_b} ((abac)^{n+1}a, aba),\tag{5.24}$$
$$((acab)^n ac|a, ac|aba) \vdash_{r_c} ((acab)^{n+1}a, aca),\tag{5.25}$$
$$((acab)^n|a, ab|aca) \vdash_{r_b} ((acab)^n aca, aba),\tag{5.26}$$
$$((abac)^n|a, ac|aba) \vdash_{r_c} ((abac)^n aba, aca) .\tag{5.27}$$

Thus, the language of the splicing system S contains the language described by the regular expression

$$(abac)^+ a \cup (acab)^* aca \cup (abac)^* aba \cup (acab)^+ a .\tag{5.28}$$

The converse inclusion also holds, since the splicing operations only lead to strings described in the above derivations. $\Diamond$

Given a splicing system $S = (\Sigma, V, R, A)$ with $\Gamma = \Sigma \cup V$, a *multiplicity* is associated with each string in Γ^*. This multiplicity is a non-negative integer indicating how often the string is available in S. At the beginning of a derivation, each axiom is assigned a multiplicity. In each application of a splicing rule $(u, v) \vdash_r (x, y)$, the multiplicities of the strings u and v are decremented by 1, while the multiplicities of the strings x and y are incremented by 1. A string with multiplicity 0 is no longer available in S and cannot be used for splicing.

5.4.2 Recursively Enumerable Splicing Systems

It will be shown that the generative power of splicing systems equals that of Turing machines. The proof follows the work of G. Păun and coworkers (1999).

Theorem 5.29. *The set of type-0 languages equals the set of languages generated by splicing systems.*

Proof. Given a type-0 grammar $G = (\Sigma, V, P, S)$ in Kuroda normal form (i.e., the rules in P are of the form (u, v) with $1 \leq |u| \leq 2$ and $0 \leq |v| \leq 2$, $u \neq v$). Put $\Gamma = \Sigma \cup V$. Define the splicing system $S = (\Sigma, V', R, A)$, where

$$V' = V \cup \{X_1, X_2, Y, Z_1, Z_2\} \cup \{(r), [r] \mid r \in P\},$$

the multiset A contains the string $w_0 = X_1^2 Y S X_2^2$ with multiplicity 1 and the following strings with infinite multiplicity:

$$w_r = (r)v[r], \quad r = (u, v) \in P,$$
$$w_\alpha = Z_1 \alpha Y Z_2, \quad \alpha \in \Gamma,$$
$$w_\alpha' = Z_1 Y \alpha Z_2, \quad \alpha \in \Gamma,$$
$$w_t = YY,$$

and the set R holds the following splicing rules:

1. $\delta_1 \delta_2 Y u \# \beta_1 \beta_2 \$(r)v\#[r]$ where $r = (u, v) \in P$,
                                         $\beta_1, \beta_2 \in \Gamma \cup \{X_2\}$, $\delta_1, \delta_2 \in \Gamma \cup \{X_1\}$,
2. $Y \# u[r] \$(r) \# v\alpha$ where $r = (u, v) \in P$, $\alpha \in \Gamma \cup \{X_2\}$,
3. $\delta_1 \delta_2 Y \alpha \# \beta_1 \beta_2 \$ Z_1 \alpha Y \# Z_2$ where $\alpha \in \Gamma$, $\beta_1, \beta_2 \in \Gamma \cup \{X_2\}$,
                                         $\delta_1, \delta_2 \in \Gamma \cup \{X_1\}$,
4. $\delta \# Y \alpha Z_2 \$ Z_1 \# \alpha Y \beta$ where $\alpha \in \Gamma$, $\delta \in \Gamma \cup \{X_1\}$,
                                         $\beta \in \Gamma \cup \{X_2\}$,
5. $\delta \alpha Y \# \beta_1 \beta_2 \beta_3 \$ Z_1 Y \alpha \# Z_2$ where $\alpha \in \Gamma$, $\beta_1 \in \Gamma$,
                                         $\beta_2, \beta_3 \in \Gamma \cup \{X_2\}$, $\delta \in \Gamma \cup \{X_1\}$,
6. $\delta \# \alpha Y Z_2 \$ Z_1 \# Y \alpha \beta$ where $\alpha \in \Gamma$, $\delta \in \Gamma \cup \{X_1\}$,
                                         $\beta \in \Gamma \cup \{X_2\}$
7. $\# Y Y \$ X_1^2 Y \# w$ where $w \in \{X_2^2\} \cup V\{X_2^2\} \cup V^2\{X_2\} \cup V^3$,
8. $\# X_2^2 \$ Y^3 \#$.

The rules in the groups 1 and 2 simulate rules in P in the presence of Y. The rules in the groups 3 and 4 move Y to the right, and the rules in the groups 5 and 6 move Y to the left. At any time, there are two occurrences of X_1 at the beginning of a string and two occurrences of X_2 at the end of a string. The rules in the groups 1, 3, and 5 separate strings of the form $X_1^2 z X_2^2$ into strings $X_1^2 z_1$ and $z_2 X_2^2$, while the rules in the groups 2, 4, and 6 bring these strings together, leading to a string of the form $X_1^2 z' X_2^2$. The rules in the groups 7 and 8 remove the auxiliary symbols X_1, X_2, and Y. If the remaining string is terminal, then it belongs to $L(G)$. Induction on the length of the derivation can be used to verify that each derivation in G can be simulated by S. Hence, $L(G) \subseteq L(S)$.

Conversely, the splicing system starts with the main axiom $w_0 = X_1^2 Y S X_2^2$, because it is not possible to splice two of the axioms w_r, w_α, w_α', and w_t. By induction, assume that we have derived a string $X_1^2 w_1 Y w_2 X_2^2$ with multiplicity 1. If w_2 starts with the left hand side of a rule in P, then we can apply to it a rule of group 1. If so, the string is $X_1^2 w_1 Y u w_3 X_2^2$ for some rule $r = (u, v) \in P$. By using the axiom $w_r = (r)v[r]$, we obtain

$$(X_1^2 w_1 Y u | w_3 X_2^2, (r)v|[r]) \vdash (X_1^2 w_1 Y u[r], |(r)vw_3 X_2^2) \;.$$

Here, no rule from the groups 1 and 3 to 8 can be applied to the derived strings, because no string containing Y^3 is derived. The rule $Y\#u[r]\$(r)\#v\alpha$ from group 2 can be applied leading to

$$(X_1^2 w_1 Y | u[r], (r)|vw_3 X_2^2) \vdash (X_1^2 w_1 Y vw_3 X_2^2, (r)u[r]) \;.$$

The string $(r)u[r]$ cannot be spliced, since in the rule $r = (u, v)$ it was assumed that $u \neq v$. The multiplicities of $X_1^2 w_1 Y u[r]$ and $(r)vw_3 X_1^2$ are 0, while the multiplicity of $X_1^2 w_1 Y vw_3 X_2^2$ is 1. Thus we derive $X_1^2 w_1 Y vw_3 X_2^2$ from $X_1^2 w_1 Y uw_3 X_2^2$, both having multiplicity 1, which corresponds to using the rule $r = (u, v)$ in P.

Moreover, at any time, there is only one string containing X_1^2 and only one string containing X_2^2. If a rule in group 3 is applied to $X_1^2 w_1 Y \alpha w_3 X_2^2$, then we derive

$$(X_1^2 w_1 Y \alpha | w_3 X_2^2, Z_1 \alpha Y | Z_2) \vdash (X_1^2 w_1 Y \alpha Z_2, Z_1 \alpha Y w_3 X_2^2) \;.$$

Here, no rules from the groups 1 to 3 and 5 to 8 can be applied to the derived strings. By using a rule from group 4, we obtain

$$(X_1^2 w_1 | Y \alpha Z_2, Z_1 | \alpha Y w_3 X_2^2) \vdash (X_1^2 w_1 \alpha Y w_3 X_2^2, Z_1 Y \alpha Z_2) \;.$$

The first resulting string $X_1^2 w_1 \alpha Y w_3 X_2^2$ replaces the string $X_1^2 w_1 Y \alpha w_3 X_2^2$, which now has multiplicity 0, and thus Y interchanges with α. The second resulting string is an axiom.

Similarly, one can see that using a rule from group 5 must be followed by using the corresponding rule from group 6, which results in interchanging Y with its left-hand neighbor.

Therefore, at each time, we have either the string $X_1^2 w_1 Y w_2 X_2^2$ or the strings $X_1^2 z_1$ and $z_2 X_2^2$, each with multiplicity 1. Only in the first case, provided $w_1 = \epsilon$, $X_1^2 Y$ can be removed by using a rule from group 7. Then we can also remove X_2^2 by using a rule from group 8. This is the only way to remove these non-terminal symbols. If the obtained string is not terminal, then it cannot be further processed, because it does not contain the symbol Y. Consequently, we can only simulate derivations in G and move Y freely in the string of multiplicity 1. Hence, $L(S) \subseteq L(G)$. $\square$

5.4.3 Universal Splicing Systems

It will be shown that programmable computers based on splicing can be built. A universal splicing system behaves as any splicing system S, when an encoding of S is introduced into the set of axioms of the universal system.

Theorem 5.30. *For each alphabet Σ, there exists a splicing system that is universal for the class of splicing systems with terminal alphabet Σ.*

Proof. For the class of type-0 grammars with terminal alphabet Σ there is a universal grammar, that is, a grammar $G_U = (\Sigma, V_U, P_U, -)$ so that for any type-0 grammar $G = (\Sigma, V, P, S)$ there is a string $S(G) \in (\Sigma \cup V_U)^*$, the *code* of G, so that for the grammar $G'_U = (\Sigma, V_U, P_U, S(G))$, we have that $L(G'_U) = L(G)$. This holds in terms of the existence of universal Turing machines and the relationship between Turing machines and type-0 grammars.

Let $G_U = (\Sigma, V_U, P_U, -)$ be a universal type-0 grammar. Following the proof in Theorem 5.29, we obtain a splicing system $S_1 = (\Sigma, V_1, R_1, A_1)$, where the axiom w_0 (with multiplicity 1) is not considered and all other axioms (with infinite multiplicity) are fixed.

First, we pass from the splicing system S_1 to the splicing system $S_2 = (\Sigma, V_2, R_2, A_2)$, which contains only one axiom (with infinite multiplicity)

$$Z = d_1 cw_r cw_\alpha cw'_\alpha cw_t cd_2 \ ,$$

where d_1, d_2, and c are newly introduced non-terminals. Moreover,

$$R_2 = R_1 \cup \{\#c\$d_2\#, \#d_1\$c\#, \} \ .$$

Thus, the string Z can be employed for cutting each original axiom. From this it follows that $L(S_2) = L(S_1)$.

Second, we pass from the splicing system S_2 to the splicing system $S_U = (\Sigma, \{c_1, c_2\}, R_U, A_U)$. For this, encode each non-terminal symbol in $V_2 = \{Z_1, \ldots, Z_n\}$ through a string over $\{c_1, c_2\}$ by using the homomorphism

$$\phi(Z_i) = c_1 c_2^i c_1, \quad 1 \leq i \leq n,$$
$$\phi(a) = a, \quad a \in \Sigma.$$

Then we put

$$A_U = \phi(A_2),$$
$$R_U = \{\phi(u_1)\#\phi(u_2)\$\phi(u_3)\#\phi(u_4) \mid u_1\#u_2\$u_3\#u_4 \in R_2\} \ .$$

It follows that $L(S_U) = L(S_2)$, since the strings $c_1 c_2^i c_1$, $1 \leq i \leq n$, are never broken by splicing and thus behave in the same way as the corresponding non-terminals in S_2.

Claim that the splicing system S_U is universal. Indeed, let $S = (\Sigma, V, R, A)$ be a splicing system. It follows from Church's thesis that there is a type-0 grammar $G = (\Sigma, V', P, S)$ so that $L(S) = L(G)$. Take the code of G, $w(G)$, consider the string

$$w'_0 = X_1^2 Y w(G) X_2^2 \, ,$$

apply the homomorphism ϕ to this string, and let $w(S)$ denote the resulting string over $\Sigma \cup \{c_1, c_2\}$. The splicing system $S'_U = (\Sigma, \{c_1, c_2\}, R_U, A_U \cup \{w(S)\})$, with the axiom $w(S)$ having multiplicity 1, has the desired property $L(S'_U) = L(S)$.   □

5.4.4 Recombinant Systems

Recombinases are enzymes that recombine DNA between specific recognition sequences. For instance, the commercially available AttSite^{TM} recombinases from RheoGene Inc., are gene-targeting enzymes that catalyze insertion, deletion or inversion of DNA at specific locations within the genome. These locations are known as attP and attB sites. Recombination is irreversible because the new sites created during recombination, attL and attR, are not recognized by these recombinases. AttSite^{TM} recombinases have robust activity in a wide range of hosts, including human stem cells and plant cells.

AttSite^{TM} recombinases give rise to three formal operations. For this, let Σ be an alphabet. Two strings over Σ are called *cyclically equivalent* if one is a cyclic permutation of the other. That is, two strings w and w' are cyclically equivalent if and only if there exist strings u and v so that $w = uv$ and $w' = vu$. This relation is an equivalence relation on Σ^*, and each equivalence class is termed a *circular string*. For instance, the string $w = abba$ belongs to the equivalence class $\{abba, aabb, baab, bbaa\}$. In the following, $\bullet w$ stands for any of the cyclic permutations of the characters in w, and $\Sigma^\bullet$ denotes the set of all circular strings over Σ.

First, *integration* (Fig. 5.11) inserts a plasmidal strand (circular DNA) into a chromosomal strand (linear DNA),

$$\left. \begin{array}{c} uBw \\ \bullet vP \end{array} \right\} \Rightarrow uLvRw \, . \tag{5.29}$$

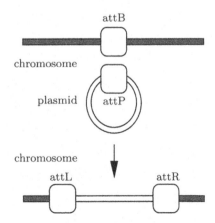

Fig. 5.11 Integration.

Fig. 5.12 Exchange.

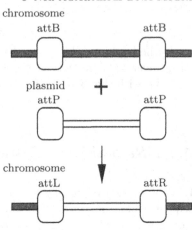

Second, *exchange* (Fig. 5.12) replaces a chromosomal substrand with a plasmidal strand,

$$\left.\begin{array}{c} uBvBw \\ \bullet v'P \end{array}\right\} \Rightarrow uLv'Rw . \tag{5.30}$$

Third, *deletion* (Fig. 5.13) deletes a substrand from a chromosome and the deleted strand forms a plasmid. This operation is the converse of integration,

$$uBvPw \Rightarrow \left\{\begin{array}{l} uLw, \\ \bullet vR . \end{array}\right. \tag{5.31}$$

More generally, a *recombinant system* is a quadruple $R = (\Sigma, V, P, S)$ consisting of a finite set of *terminal symbols* Σ, a finite set of *non-terminal symbols* V with $\Sigma \cap V = \emptyset$, a finite set of *production rules* P providing integration, exchange, and deletion, and a *start symbol* $S \in V$.

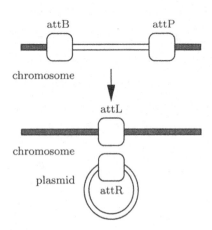

Fig. 5.13 Deletion.

Integration is of the form $(A, \bullet yB) \to_R CyD$, exchange has the shape $(AwA, \bullet yB) \to_R CyD$, and deletion has the form $AyB \to_R (C, \bullet yD)$, where $A, B, C, D \in V$ and $w, y \in (\Sigma \cup V)^*$.

Let $u, v \in (\Sigma \cup V)^*$. Write $u \Rightarrow_R v$ if v is derived from u by integration, exchange, or deletion. *Integration* means $(A, \bullet yB) \to_R CyD \in P$ and

$$u = xAz, \tag{5.32}$$
$$v = xCyDz, \qquad x, z \in (\Sigma \cup V)^*.$$

Exchange stands for $(AwA, \bullet yB) \to_R CyD \in P$ and

$$u = xAwAz, \tag{5.33}$$
$$v = xCyDz, \qquad w, x, z \in (\Sigma \cup V)^*.$$

Deletion signifies $AyB \to_R (C, \bullet yD) \in P$ and

$$u = xAyBz, \tag{5.34}$$
$$v = xCz, \qquad x, z \in (\Sigma \cup V)^*.$$

Additional production rules of the form $C \to_R \epsilon$ are required, where C is a non-terminal symbol occurring only on the right hand side in the production rules of integration, exchange, and deletion. Let $\Rightarrow_R^*$ denote the reflexive and transitive closure of $\Rightarrow_R$. The *language* generated by R is given by

$$L(R) = \{w \in \Sigma^* \mid S \Rightarrow_R^* w\} . \tag{5.35}$$

Theorem 5.31. *For each type-2 grammar G there is a recombinant system R, which only makes use of integration, so that $L(G) = L(R)$, and vice versa.*

Proof. Let $G = (\Sigma, V, P, S)$ be a type-2 grammar. For each rule $A \to_G w$ define the integration rule $(A, \bullet wB) \to_R CwD$, where C and D only occur on the right hand side in the production rules. The corresponding recombinant system $R = (\Sigma, V, P', S)$ satisfies $L(R) = L(G)$. This assertion can be reversed. $\qquad\square$

Example 5.32. Consider the type-2 grammar G in Example 2.43. The corresponding recombinant system R is given by $\Sigma = \{a, b\}$, $V = \{S, B, C, D\}$, start symbol S, and the production rules $(S, \bullet aSbB) \to CaSbB$, $(S, \bullet abB) \to CabD$, $C \to \epsilon$, and $D \to \epsilon$. For instance, the string $a^3 b^3$ is derived in G and R as follows:

$$S \Rightarrow_G aSb \Rightarrow_G aaSbb \Rightarrow_G aaabbb,$$
$$S \Rightarrow_R CaSbD \Rightarrow_R CaCaSbDbD \Rightarrow_R CaCaCabDbDbD \Rightarrow_R^* aaabbb .$$

$\diamondsuit$

Theorem 5.33. *The set of type-0 languages equals the set of languages generated by recombinant systems.*

Proof. Let $G = (V, \Sigma, P, S)$ be a type-0 grammar. Three cases are distinguished. First, for each rule $S \to_G w$ define the integration rule $(S, \bullet wB) \to_R CwD$. Second, for each rule $v \to_G w$ with $w \neq \epsilon$ define the exchange rule $(AvA, \bullet wB) \to_R CwD$. Third, for each rule $v \to_G \epsilon$ define the deletion rule $AvA \to_R (C, \bullet vD)$. The non-terminals C and D are chosen so that they occur only on the right hand sides in the production rules. The corresponding recombinant system $R = (\Sigma, V, P', S)$ satisfies $L(R) = L(G)$. The converse follows by using Church's thesis.   □

Example 5.34. Consider the type-1 grammar G in Example 2.41. The corresponding recombinant system R over the alphabet $\{a, b, c\}$ has the production rules $(S, \bullet aBScX) \to CaBScD$, $(S, \bullet abcX) \to CabcD$, $(XBaX, \bullet aBY) \to CaBD$, $(XBbX, \bullet bbY) \to CbbD$, $C \to \epsilon$, and $D \to \epsilon$, and start symbol S. For instance, the word $a^2b^2c^2$ is derived in G and R as follows:

$$S \Rightarrow_G aBSc \Rightarrow_G aBabcc \Rightarrow_G aaBbcc \Rightarrow_G aabbcc,$$
$$S \Rightarrow_R CaBScD \Rightarrow_R CaBCabcDcD \stackrel{*}{\Rightarrow}_R aBaabcc \Rightarrow_R aCaBDbcc$$
$$\Rightarrow_R aaBbcc \Rightarrow_R aaCbbDcc \stackrel{*}{\Rightarrow}_R aabbcc .$$

Concluding Remarks

The first generation of DNA computing mainly aimed to tackle hard computational problems. For this, new paradigms of computing were devised that basically filter a large search space in a parallel manner. However, these paradigms do not provide new insights into computational complexity theory. Problems which grow exponentially with the problem size on a conventional computer still grow exponentially with the problem size on a DNA machine. For very large problem instances, the amount of the DNA required is too large to be practical. On the other hand, the DNA algorithms of the first generation may be suitable to run on dedicated hardware if combined with appropriate heuristics that provide the data objects in question.

References

1. Adleman L (1994) Molecular computation of solutions of combinatorial problems. Science 266:1021–1023
2. Adleman L (1996) On constructing a molecular computer. DIMACS 27:1–21
3. Amos M (2005) Theoretical and experimental DNA computation. Springer, Berlin Heidelberg

4. Bach E, Condon A, Glaser E, Tanguay C (1996) DNA models and algorithms for NP-complete problems. Proc 11th Ann IEEE Conf Comp Complex, Philadelphia, 290–299
5. Braich RS, Chelyapov N, Johnson C, Rothemumd PWK, Adleman L (2002) Solution of a 20-variable 3-sat problem on a DNA computer. Science 296:499–502
6. Faulhammer D, Cukras AR, Lipton RJ, Landweber LF (2000) Molecular computation: RNA solutions to chess problems. PNAS 97:1385–1389
7. Feller W (1968) An introduction to probability theory and its applications. Wiley, New York
8. Feynman RP (1961) Miniaturization. In: Gilbert DH (ed.). Reinhold, New York
9. Freund R, Kari L, Păun G (1999) DNA computing based on splicing: the existence of universal computers. Theory Comp Systems 32:69–112
10. Gibbons A, Amos M, Hodgson D (1996) Models of DNA computation. LNCS 1113:18–36
11. Guo M, Chang WL, Ho M, Lu J, Cao J (2005) Is optimal solution of every NP-complete or NP-hard problem determined from its characteristic for DNA-based computing? Biosystems 80:71–82
12. Head T (1987) Formal language theory and DNA: an analysis of the generative capacity of specific recombinant behaviors. Bull Math Biol 47:737–759
13. Henkel CV, Rozenberg G, Spaink H (2005) Application to mismatch detection methods in DNA computing. LNCS 3384:159–168
14. Lipton RJ (1995) DNA solution of hard combinatorial problems. Science 268: 542–545
15. Liu Q, Wang L, Frutos AG, Condon AE, Corn RM, Smith LM (2000) A surface-based approach to DNA computing. Nature 403:175–179
16. Liu Q, Wang L, Frutos AG, Condon AE, Corn RM, Smith LM (2000) DNA computing on surfaces. Nature 403:175–179
17. Martinez-Perez I (2007) Biomolecular computing models for graph problems and finite state automata. Ph.D. thesis Hamburg Univ Tech
18. Ouyang Q, Kaplan PD, Liu S, Libchaber A (1997) DNA solution of the maximal clique problem. Science 278:446–449
19. Păun G, Rozenberg G, Salomaa A (1998) DNA computing: new computing paradigms. Springer, New York
20. RheoGene (2005) Market Wire
21. Roweis S, Winfree E, Burgoyne R, Chelyapov N, Goodman M, Rothemund P, Adleman L (1996) A sticker based architecture for DNA computation. In: Baum EB (ed.) DNA Based Computers 1–27
22. Rozenberg G, Spaink H (2003) DNA computing by blocking. Theoret Comp Sci 292:653–665
23. Sakakibara Y, Suyama A (2000) Intelligent DNA chips: logical operation of gene expression proviles on DNA computers. Genome Inform 11:33–42
24. Scharrenberg O (2007) Programming of stickers machines. Project Work, Hamburg Univ Tech
25. Zimmermann KH (2002) On applying molecular computation to binary linear codes. IEEE Trans Inform Theory 48:505–510
26. Zimmermann KH (2002) Efficient DNA sticker algorithms for NP-complete graph problems. Comp Phys Comm 114:297–309

Chapter 6
Autonomous DNA Models

Abstract The second generation of DNA computing focusses on models that are molecular-scale, autonomous, and partially programmable. The computations are essentially driven by the self-assembly of DNA molecules and are modulated by DNA-manipulating enzymes. This chapter addresses basic autonomous DNA models emphasizing tile assembly, finite state automata, Turing machines, neural networks, and switching circuits.

6.1 Algorithmic Self-Assembly

The idea of algorithmic self-assembly arose from the combination of DNA computing, tiling theory, and DNA nanotechnology. Algorithmically self-assembled structures span a wide range between maximally simple structures (crystals) and arbitrarily complex information processing tilings (supramolecular complexes). Algorithmic self-assembly is amenable to experimental investigations allowing the understanding of involved physical phenomena. This understanding may eventually result in new nanostructure materials and devices.

6.1.1 Self-Assembly

Algorithmic self-assembly was first explored by E. Winfree in the late 1990s. We follow his remarkable thesis (1998) in order to introduce a mathematical model of DNA self-assembly. This model is based on three basic operations: annealing, ligation, and denaturation. A physical system realizing DNA self-assembly may work as follows:

- Synthesize several DNA molecules.
- Mix the DNA together in solution.

Z. Ignatova et al., *DNA Computing Models*,
DOI: 10.1007/978-0-387-73637-2_6, © Springer Science+Business Media, LLC 2008

- Heat up the solution and slowly cool it down, allowing complexes of double-stranded DNA to form.
- Ligate adjacent strands.
- Denaturate the DNA again.
- Determine what single-stranded DNA molecules are present in the solution.

The outcome of such a physical self-assembly process is hard to predict. Therefore, we simplify the mathematical model of DNA self-assembly by introducing the following constraints:

- *Constant temperature:* The number of base pairs required for the stability of DNA molecules remains invariant during self-assembly. We consider neither annealing at high temperature allowing only long regions to hybridize nor annealing at low temperature stimulating the hybridization of short regions.
- *Perfect complementarity:* Annealing only takes place between DNA strands with perfect Watson-Crick complementarity, while the annealing of mismatched DNA strands that creates unusual structures such as bubbles, branched junctions or triple helices is not considered.
- *Permanent binary events:* Each self-assembly event is binary (i.e., occurs only between two DNA molecules), and permanent (i.e., joined DNA molecules never dissociate).
- *Initial DNA molecules:* Some DNA molecules are formed prior to self-assembly. These molecules include single-stranded DNA, double-stranded DNA with sticky ends, and possibly some unusual DNA structures.
- *Intramolecular events:* A self-assembled DNA molecule may interact with itself by forming secondary structures such as hairpins.
- *Single or multiple binding regions:* A binary annealing event can create either a single contiguous Watson-Crick region or multiple contiguous Watson-Crick regions.

The first four constraints are used throughout this section, while the last two contraints may vary from model to model.

6.1.2 DNA Graphs

A DNA graph represents several DNA polynucleotides bound together by Watson-Crick hybridization. Formally, a *DNA graph* is a connected graph with vertices labelled from the DNA alphabet $\Delta = \{A, C, G, T\}$ and edges labelled as either *backbone* or *base-pair*. The backbone edges are directed indicating the 5' to 3' direction, while the base-pair edges are undirected and are of the Watson-Crick type. Moreover, each vertex in a DNA graph has at most one incoming and one outgoing edge of each type. Specifically, there are ten DNA graphs with two backbone edges and two base-pair edges (Fig. 6.1).

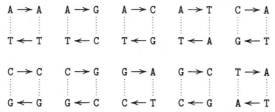

Fig. 6.1 The 10 Watson-Crick subgraphs.

DNA graphs can represent a rich variety of DNA structures, but structures such as triple helices are missing and notions of geometry and topology are lacking. However, it is possible that DNA graphs specify physically impossible structures.

Three basic operations on DNA graphs are introduced, each having a physical counterpart:

- *Annealing*: Take two DNA graphs D_1 and D_2 and provide a compound DNA graph $D_3 = D_1 + D_2$ by forming base-pair edges linking vertices in D_1 with vertices in D_2.
- *Ligation*: Consider a DNA graph D and produce a DNA graph $D' = -D$ by adding backbone edges linking vertices in D.
- *Denaturation*: Pick a DNA graph D and derive a set of DNA graphs $D^d = \{D_i\}$, each being a backbone component of D with no base-pair edges.

Example 6.1. Consider two DNA graphs with complementary sticky ends (Fig. 6.2). Annealing yields a compound DNA graph. After ligation, the DNA graph is denatured, resulting in two single strands.   ◊

Let A be a set of DNA graphs. The *language* of A is given by the set $L(A)$ of all DNA graphs that can be obtained from the DNA graphs in A by a finite number of annealing and ligation cycles. DNA annealing and ligation can produce many unusual structures in addition to the usual B-form double helix such as hairpins, three armed junctions, and double crossover (DX) molecules (Fig. 6.5).

Denaturating each DNA graph in $L(A)$ yields a set of single DNA strands that is called the *denaturation language* of A given by

$$L_d(A) = \bigcup \{D^d \mid D \in L(A)\} . \tag{6.1}$$

Single DNA strands can be translated into strings over another alphabet Σ by using a *codebook* over Σ, which assigns to each symbol a in Σ a string $C_\Sigma(a)$ over Δ. This assignment must obey the *Fano condition*, which says that no string in the codebook is a prefix of another string in the codebook. Otherwise, the assignment will not be well-defined. A DNA strand $x = a_1 \ldots a_n$ is translated into a string over Σ by scanning through the strand x from left to right so that if a_i begins a subsequence of x which matches

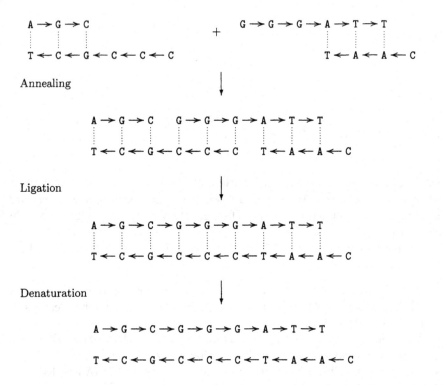

Fig. 6.2 Annealing, linkage, and denaturation of two DNA graphs.

some $C_\Sigma(a)$, then a_i is replaced by a. Otherwise, a_i is deleted and then a_{i+1} is processed. Write $C_\Sigma^{-1}(x)$ for the correspondingly translated string over Σ.

Example 6.2. Consider the codebook over $\Sigma = \{a, b\}$ given by $C_\Sigma(a) = \text{AGT}$ and $C_\Sigma(b) = \text{ACA}$. This codebook satisfies the Fano condition and maps the DNA strand $x = \text{GGGACACAGTACAT}$ to $C_\Sigma^{-1}(x) = bbab$. $\Diamond$

In this way, the denaturation language of A and the codebook C over Σ provide the *language* of A and C over Σ defined as

$$L_d(A, C) = \bigcup \{C_\Sigma^{-1}(D^d) \mid D \in L(A)\}, \tag{6.2}$$

where $C_\Sigma^{-1}(D^d) = \{C_\Sigma^{-1}(x) \mid x \in D^d\}$ is the set of strings translated from the denaturated DNA graph D^d via the codebook.

6.1.3 Linear Self-Assembly

Linear self-assembly proceeds at a constant temperature, affords perfect Wat-son-Crick complementarity, enables permanent binary events, and disallows intramolecular events. Moreover, linear self-assembly is based on binary annealing events that create single contiguous Watson-Crick regions and begins with single-stranded DNA or double-stranded DNA with sticky ends.

Theorem 6.3. *Let Σ be an alphabet. For each regular language L over Σ, there exists a set A of single strands or double strands with sticky ends and a codebook C over Σ so that $L = L_d(A, C)$. For each set A of single strands and double strands with sticky ends and each codebook C over Σ, the language $L_d(A, C)$ is regular over Σ.*

Proof. Let L be a regular language over Σ. There is a regular grammar G with $L(G) = L$ so that all production rules are of the form (A, aB), (A, a) or (A, ϵ), where A and B are non-terminal symbols and a is a terminal symbol. First, design sufficiently dissimilar DNA sequences for all terminal and non-terminal symbols in G. Denote the encoding of a symbol X by X_s and let $\overline{X}_s$ denote its Watson-Crick complement. For each production rule (A, aB), design a double strand beginning with sticky end $\overline{A}_s$, followed by double strand a_s, and ending with sticky end B_s. Similarly, for each production rule (A, a), design a double strand beginning with sticky end $\overline{A}_s$ followed by double strand a_s. Moreover, for each production rule (A, ϵ), design a single strand $\overline{A}_s$. Finally, for the start symbol S design a double strand starting with a blunt end followed by sticky end S_s. These single strands and double strands with sticky ends form the initial set A. After self-assembly, the complexes in $L(A)$ correspond to derivations in G. After ligation, each DNA graph in $L(A)$ will be double-stranded without sticky ends whose sequence consists of terminal symbols interspersed with non-terminal symbols. After denaturation, the codebook given by $C_\Sigma(a) = a_s$ for each terminal symbol a will delete the non-terminal symbols and provide the language $L = L_d(A, C)$, as required.

Conversely, the same construction from a set A of single strands and double strands with sticky ends leads, via a codebook C over Σ, to a regular grammar G so that $L(G) = L_d(A, C)$. $\square$

Example 6.4. Consider the regular grammar G given by the terminal set $\Sigma = \{a, b\}$, non-terminal set $V = \{S, A\}$, start symbol S, and production rules (S, ϵ), (S, a), (S, aS), (S, bA), (A, aA), and (A, bS). The corresponding language consists of all strings over Σ with an even number of b's. Encode the grammar by the set A of partially double-stranded DNA graphs given in Figure 6.3, and take the codebook over Σ given by $C_\Sigma(a) = \text{AGT}$ and $C_\Sigma(b) = \text{ACA}$. It follows that the language $L_d(A, C)$ equals the language of G (Fig. 6.4). $\Diamond$

6.1.4 Tile Assembly

The tile assembly model introduced by E. Winfree (1998) connects the theory of tilings with structural DNA nanotechnology first studied by N. Seeman (1982).

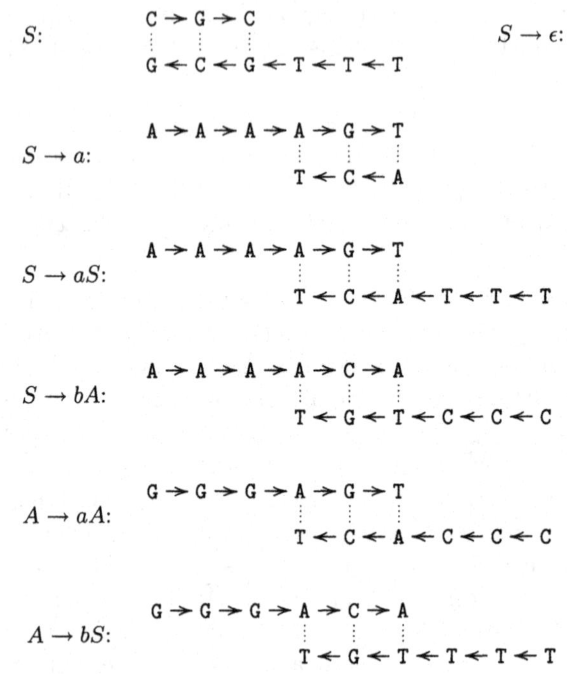

S:
$$C \rightarrow G \rightarrow C$$
$$\vdots \quad \vdots \quad \vdots$$
$$G \leftarrow C \leftarrow G \leftarrow T \leftarrow T \leftarrow T$$

$S \rightarrow \epsilon$: $A \leftarrow A \leftarrow A$

$S \rightarrow a$:
$$A \rightarrow A \rightarrow A \rightarrow A \rightarrow G \rightarrow T$$
$$\vdots \quad \vdots \quad \vdots$$
$$T \leftarrow C \leftarrow A$$

$S \rightarrow aS$:
$$A \rightarrow A \rightarrow A \rightarrow A \rightarrow G \rightarrow T$$
$$\vdots \quad \vdots \quad \vdots$$
$$T \leftarrow C \leftarrow A \leftarrow T \leftarrow T \leftarrow T$$

$S \rightarrow bA$:
$$A \rightarrow A \rightarrow A \rightarrow A \rightarrow C \rightarrow A$$
$$\vdots \quad \vdots \quad \vdots$$
$$T \leftarrow G \leftarrow T \leftarrow C \leftarrow C \leftarrow C$$

$A \rightarrow aA$:
$$G \rightarrow G \rightarrow G \rightarrow A \rightarrow G \rightarrow T$$
$$\vdots \quad \vdots \quad \vdots$$
$$T \leftarrow C \leftarrow A \leftarrow C \leftarrow C \leftarrow C$$

$A \rightarrow bS$:
$$G \rightarrow G \rightarrow G \rightarrow A \rightarrow C \rightarrow A$$
$$\vdots \quad \vdots \quad \vdots$$
$$T \leftarrow G \leftarrow T \leftarrow T \leftarrow T \leftarrow T$$

Fig. 6.3 Encoding of regular grammar.

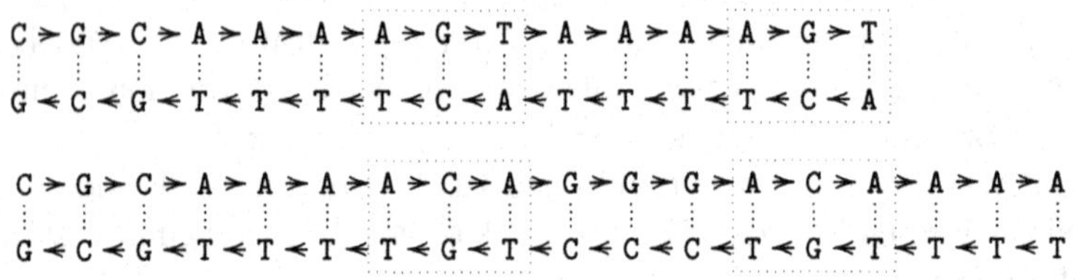

$$C \succ G \succ C \succ A \succ A \succ A \succ A \succ G \succ T \succ A \succ A \succ A \succ A \succ G \succ T$$
$$\vdots \quad \vdots \quad \vdots \quad \vdots \quad \vdots \quad \vdots \quad \vdots \quad \vdots$$
$$G \prec C \prec G \prec T \prec T \prec T \prec T \prec C \prec A \prec T \prec T \prec T \prec T \prec C \prec A$$

$$C \succ G \succ C \succ A \succ A \succ A \succ A \succ C \succ A \succ G \succ G \succ G \succ A \succ C \succ A \succ A \succ A \succ A$$
$$\vdots \quad \vdots \quad \vdots \quad \vdots \quad \vdots \quad \vdots \quad \vdots \quad \vdots \quad \vdots$$
$$G \prec C \prec G \prec T \prec T \prec T \prec T \prec G \prec T \prec C \prec C \prec C \prec T \prec G \prec T \prec T \prec T \prec T$$

Fig. 6.4 Linear self-assembly of strings aa and bb by using the DNA strands shown in Figure 6.3. The framed entries provide the symbols via the codebook.

Double Crossover Molecules

Combinatorial tiles can be realized by *double-crossover (DX) molecules* introduced by T.-J. Fu and N. Seeman (1993). DX molecules are DNA structures with four sticky ends containing two crossovers that connect two helical domains. There are several types of DX molecules, differentiated by the number of double helical half-turns, even (E) or odd (O), and by the relative orientation of their helix axes, parallel (P) or anti-parallel (A) (Fig. 6.5). DX molecules are analogues of intermediates in meiosis that consist of two side-by-side double-stranded helices linked at two crossover junctions. In particular, anti-parallel DX molecules of types DAO and DAE exhibit rigidity,

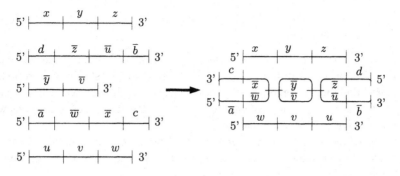

Fig. 6.5 Construction of DAE.

making them suitable for use in the self-assembly of periodic matter. The four sticky ends of DX molecules can be encoded accordingly so that they correspond to the labels on the four sides of the Wang tiles (Fig. 6.6). This allows in principle to encode any finite set of Wang tiles by DX molecules.

Self-Assembly of DX Molecules

The self-assembly of DX molecules needs to be controlled so that the arrangement of DX molecules results in a tiling. For this, a specific self-assembly model called *tile assembly model* is considered. This model proceeds at a constant temperature, affords perfect Watson-Crick complementarity of the DX molecule's sticky ends, enables permanent binary events, disallows intramolecular events, and involves multiple binding regions.

   In the tile assembly model, an unattached DX molecule may anneal to a two-dimensional lattice of DX molecules if there is a free slot in the lattice into which the molecule will fit and the molecule's sticky ends can anneal to free sticky ends in the lattice. It is crucial that a DX molecule that matches one side in a slot but not the other side will not anneal to the lattice. Under appropriate conditions, DX molecules binding to only one side in a slot soon dissociate, while fully matching DX molecules permanently bind. Therefore, slot filling is considered as one single permanent binary event involving two binding regions and the temperature needs to be chosen so that binding of

Fig. 6.6 Representation of DAE as a tile.

single sites will not occur. The tile assembly model makes use of the following
additional assumptions:

- *Binding strength:* The DNA annealing strength of DX molecules primarily
  depends on the number of base pairs so that longer sticky ends represent
  edge labels with higher strength and vice versa. In this way, the strength
  of edge labels can be implemented by designing sticky end sequences with
  specific energetics of annealing.
- *Cooperative binding strength:* The binding of DX molecules where two
  sticky end sequences anneal is cooperative in the sense that the binding
  strength of sticky-end annealing is additive.
- *Strength parameter:* There is a physical parameter T that determines
  the strength required for association of molecular tiles. In this way, DX
  molecules will only stick to a growing crystal if they bind with a total
  strength greater than or equal to the threshold T, while DX molecules
  with a weaker strength immediately fall off.

The temperature can serve as a strength parameter, because sticky ends
bind more strongly at low temperatures, while at higher temperatures the
sticky ends need to be longer for stable addition. Let T_0, T_1, and T_2 be
the melting temperatures for a DX molecule fitting into a lattice slot where
respectively 0, 1, and 2 of the sticky end pairings match. Since higher binding
strength is directly proportional to higher melting temperature, it follows that
$T_0 < T_1 < T_2$. Equivalently, the discrete values 0, 1, and 2 can be used to
indicate the respective temperatures.

A computation in the tile assembly model starts from an arrangement of
seed tiles and proceeds in a controlled manner by annealing, linkage, and
denaturation of DX molecules.

Example 6.5. Consider the Pascal triangle and its modulo 2 variant termed
Sierpinsky triangle:

$$
\begin{array}{ccccccccc}
 & & & 1 & & & & & \\
 & & 1 & & 1 & & & & \\
 & 1 & & 2 & & 1 & & & \\
1 & & 3 & & 3 & & 1 & & \\
 & & & \cdots & & & & &
\end{array}
\qquad \xrightarrow{\;\text{mod } 2\;} \qquad
\begin{array}{ccccccccc}
 & & & 1 & & & & & \\
 & & 1 & & 1 & & & & \\
 & 1 & & 0 & & 1 & & & \\
1 & & 1 & & 1 & & 1 & & \\
 & & & \cdots & & & & &
\end{array}
$$

1   4   6   4   1   1   0   0   0   1

In the Pascal triangle, the kth entry in the nth row is given by the binomial
number $\binom{n}{k}$, which can be calculated from its above left and right neighbors by
the formula (5.3). Thus, the kth entry in the nth row of the Sierpinsky triangle
is simply computed by adding its above left and right neighbors modulo 2.
This observation provides the rule tiles in which all edges have strength 1
(Fig. 6.7). Moreover, three types of boundary tiles are defined (Fig. 6.8).

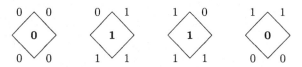

Fig. 6.7 Rule tile set for the Sierpinski triangle with all edges having strength 1. The values of the left and right neighbors are given above, and their sum modulo 2 is provided inside the tile and propagated below to the left and to the right.

Fig. 6.8 Boundary tile set for the Sierpinsky triangle. Double edges have strength 2, thin edges strength 1, and thick edges strength 0.

At temperature $T = 0$, every possible monomer addition is stable, and thus random aggregates are produced. At temperature $T = 1$, at least one edge must match for an addition to be stable, but the arrangement of tiles within an aggregate depends upon the sequence of addition. At temperature $T = 2$, self-assembly starting from an arrangement of seed tiles results in a unique tiling corresponding to the Sierpinsky triangle (Fig. 6.9). At temperature $T = 3$, no aggregates are formed.   ◇

Tile assembly can be utilized to generate words from the language of a grammar.

Example 6.6. Consider the regular grammar G with terminal set $\{a\}$, nonterminal set $\{S, A\}$, start symbol S, and production rules (S, A), (A, a), and (A, Aa). The production rules of G can be realized by the DX molecules illustrated in Figure 6.10. Each such molecule has two or three sticky ends

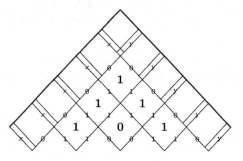

Fig. 6.9 Tiling corresponding to the Sierpinsky triangle for temperature $T = 2$. The arrangement of seed tiles consists of the four upper tiles. The upper middle tile indicating 1 is only available in the seed arrangement.

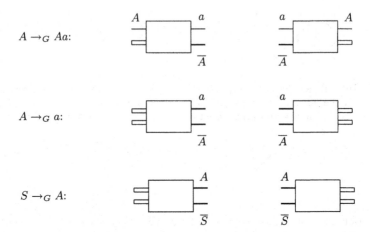

$A \to_G Aa$:

$A \to_G a$:

$S \to_G A$:

Fig. 6.10 DX molecules with two or three sticky ends encoding the production rules of a regular grammar. Sticky ends are indicated by single lines while closed loops are given by closed double lines. Each sticky end is labelled by a terminal or non-terminal, where the Watson-Crick complement of a non-terminal X is denoted by $\overline{X}$. Sticky ends indicated by thick lines have strength 2, while the remaining have strength 1.

encoding the left and right hand sides of the associated production rule. The seed tile is given by the production involving the start symbol S. Self-assembly at temperature $T = 2$ eventually results in a lattice providing a derivation given by the grammar G.   ◇

Theorem 6.7. *The tile assembly model is universal.*

Proof. Given a one-dimensional BCA C. Claim that C can be simulated by the tile assembly model. Indeed, for each transition rule $(u, v) \to_C (x, y)$ in C create a DX molecule whose sticky ends on one helix are the Watson-Crick complements of u and v, and on the other helix are x and y (Fig. 6.11). These DX molecules are added to the solution, which contains an arrangement of seed tiles encoding the initial state of the automaton C (Fig. 6.12). All sticky ends in the DX molecules are assumed to have strength 1. Then the computation results in a unique tiling encoding the space-time history of the BCA.   □

The simulation of a one-dimensional BCA by the tile assembly model poses two questions. The first question addresses the termination of the simulation.

Fig. 6.11 DX molecule for the BCA transition rule $(u, v) \to_C (x, y)$.

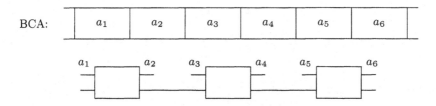

Fig. 6.12 Initial state of BCA and corresponding arrangement of seed tiles.

For this, suppose a computation of the one-dimensional BCA terminates when a special *halting symbol* appears for the first time anywhere in a cell. The halting symbol can be designed so that it corresponds to a special sticky end motif. This motif can be chosen as the recognition site for a binding protein, which could subsequently catalyze a phosphorescent reaction indicating termination. However, when the halting symbol first appears in the BCA, the other cells are not aware that the computation is done. To this end, the cells can be synchronized so that they enter a special state after the same number of steps, and so terminate the computation at the same time. This problem is termed *firing squad problem* and was first posed by J. Myhill in the late 1950s. After the phosphorescent reaction, E. Winfree suggested designing linear pieces of DNA that fit into the gaps at the final level of the lattice so that it cannot grow further.

The second question concerns the output of the simulation result. To this end, the level in the lattice containing the halting symbol needs to be analyzed. Here, E. Winfree proposed adding resolvase to break all crossover junctions, then heating the solution in order to separate the strands, and using affinity purification to extract the strand containing the halting motif.

Tile Assembly in Boolean Arrays

The Sierpinsky triangle can be specified by a two-dimensional Boolean array. Now the tile assembly of general $m \times n$ Boolean arrays is addressed. For this, assume that each tile $T_0(i,j)$, $0 \le i \le m - 1$, $0 \le j \le n - 1$, has the form of a square and each side or pad of the square is associated with a Boolean variable. The tile $T_0(i,j)$ has two input pads and two output pads (Fig. 6.13). The input pads are given by Boolean variables $b(i,j)$ (bottom) and $r(i,j)$ (right). The output pads correspond to Boolean variables $b(i,j+1)$ (top) and $r(i+1,j)$ (left) and are defined by 2-adic Boolean functions $\oplus$ and $\ominus$ as follows,

$$b(i,j+1) = b(i,j) \oplus r(i,j) \quad \text{and} \quad r(i+1,j) = b(i,j) \ominus r(i,j) . \quad (6.3)$$

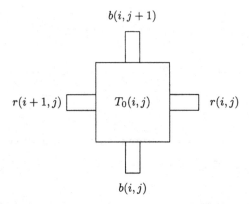

Fig. 6.13 The tile $T_0(i,j)$.

Assume that the bottom row and the left-most column of the array provide the inputs to the assembly (i.e., the Boolean variables $b(0,j)$, $0 \leq j \leq n-1$, and $r(i,0)$, $0 \leq i \leq m-1$, are predefined). Then Eq. (6.3) can be used to fully evaluate the Boolean array. This naive tile assembly model requires three tile types for assembling the initial frame and $2^2 = 4$ tile types for the interior part, as each tile type depends on two input pads.

Example 6.8. The Sierpinsky triangle can be considered as an $n \times n$ Boolean array, where the bottom row and the left most column all have 1's, and the functions $\oplus$ and $\ominus$ are both exclusive ORs. $\diamond$

A critical issue in tile assembly is the *pad mismatch rate*, which determines the size of the error-free assembly. Let ϵ be the probability of a single pad mismatch between adjacent assembling tiles so that they still stay together in equilibrium. Assume that the likelihood of a pad mismatch is independent of any other match or mismatch. This *independent error model* was studied by J.H. Reif and coworkers (2007). If an error in a pad in a tile enforces k further mismatches in the assembly in the immediate neighborhood of that tile, then the error probability reduces to ϵ^{k+1}. Indeed, if one error enforces k more errors, then the probability that the tile and its neighborhood in the assembly will stay together in the equilibrium in spite of these $k+1$ errors is ϵ^{k+1}. A key challenge in tile assembly is to make the tiles error-resilient. The basic technique to achieve this is to extend the tile complexity.

Accretive Graph Assembly Model

A more general tile assembly model was introduced by J.H. Reif and coworkers (2005–2006). First, it involves not only attractive forces but also repulsive forces. Repulsive forces occur between hydrophobic and hydrophilic tiles, or between tiles labelled by magnetic or charged pads of the same polarity. Second, the model allows the assembly of graph-like structures and not only rectangular tile structures.

A *graph assembly system* consists of a graph $G = (V, E)$, a distinguished vertex $v_0 \in V$, an edge-weight function $w : E \to \mathbb{Z}$, and a temperature $T \in \mathbb{Z}$. The graph G of a graph assembly system is *sequentially constructible* if all its vertices can be attached one by one, starting from the distinguished vertex, so that the support of the already assembled adjacent vertices exceeds the temperature. That means, there is a total ordering of the vertices, $v_0 \prec v_1 \prec \ldots \prec v_{n-1}$ beginning with the distinguished vertex, so that

$$\sum_j w(v_j v_i) > T, \quad 1 \le i \le n - 1, \tag{6.4}$$

where the sum extends over all vertices v_j adjacent to v_i with $j < i$. The system is *accretive* in the sense that already assembled components cannot be subsequently removed from the assembly.

Example 6.9. Consider the graph in Figure 6.14. Take the strength $T = 1$. The graph is sequentially constructible by the assembly order $a \prec b \prec c \prec d \prec f \prec g \prec h \prec i \prec e$. However, the vertex i cannot be assembled without the support of the vertex h, that is, $h \prec i$. The vertex e can be assembled if two of the vertices b, d, and f are present and h is absent. But if the vertex h is present, then the vertex e needs support from all three vertices b, d, and f. $\diamond$

Theorem 6.10 (Accretive Graph Assembly Problem). *The problem of finding an assembly ordering in a graph assembly system so that the graph is sequentially constructible is NP-hard.*

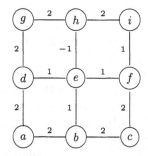

Fig. 6.14 A graph for an accretive graph assembly system.

6.2 Finite State Automaton Models

This section provides models of finite state automata that operate autonomously at the molecular scale.

6.2.1 Two-State Two-Symbol Automata

The first DNA model for finite state automata working autonomously at the molecular scale was invented by E. Shapiro and coworkers in 2001. This model is based on the DNA Turing machine model developed by P. Rothemund in 1996.

Data Representation

The DNA model of the Shapiro group implements two-state two-symbol automata. For this, let M be a finite state automaton with input alphabet $\Sigma = \{a, b\}$ and state set $S = \{s_0, s_1\}$. This automaton can have 8 possible transition rules and the programming of the automaton M amounts to selecting some of these transition rules and deciding which states are initial and final. There are $2^8 = 256$ possible selections of transition rules and three possible selections of both initial and final states (either s_0 or s_1, or both).

The implementation of these two-state two-symbol automata is composed of hardware, software, and input. The hardware consists of a mixture of restriction endonuclease FokI, ligase, and ATP. The enzyme FokI has the recognition site

$$5' - GGATG - 3'$$
$$3' - CCTAC - 5'$$

and cuts downstream from the recognition site at subsequent positions 9 and 13 leaving sticky ends, for example,

$$5' - GGATGTACGGCTCG \mid CAGCA - 3'$$
$$3' - CCTACATGCCGAGCGTCG \mid T - 5'$$

where the vertical bars indicate the cleavage site. The software comprises 8 short double-stranded molecules encoding the transition rules (Fig. 6.15). The input given by the initial state and an input string is encoded by a double-stranded molecule (Fig. 6.16). Here, the initial state is always s_0 and each input symbol is encoded by six base pairs (Fig. 6.17). The automaton also contains two output detection molecules of different lengths (Fig. 6.18). Each molecule can interact with a distinguished output molecule to form an output reporter molecule indicating a final state. Because these molecules

δ(s0, a) = s0 :
3' GGT 22 GGATGTAC CCTACATGCCGA 5'
δ(s0, a) = s1 :
3' GGT 22 GGATGACGAC CCTACTGCTGCCGA 5'
δ(s0, b) = s0 :
3' GGT 28 GGATGACG CCTACTGCGTCG 5'
δ(s0, b) = s1 :
3' GGT 15 GGATGACGAC CCTACTGCTGGTCG 5'
δ(s1, a) = s0 :
3' GGT 28 GGATGA CCTACTGACC 5'
δ(s1, b) = s0 :
3' GGT 21 GGATGG CCTACCGCGT 5'
δ(s1, a) = s1 :
3' GGT 15 GGATGACG CCTACTGCGACC 5'
δ(s1, b) = s1 :
3' GGT 30 GGATGACG CCTACTGCGCGGT 5'

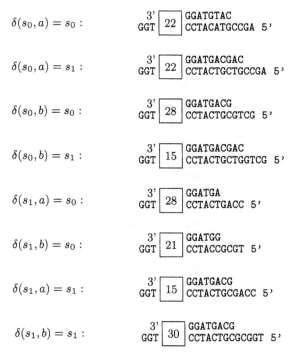

Fig. 6.15 Encoding of the transition rules in a two-symbol two-state automaton.

have different lengths, the final state can be detected by length separation such as gel electrophoresis.

Computations

A computation of Shapiro's automaton begins by mixing the hardware, software, and input together into one test tube. The computation processes the input molecule by a cascade of restriction, hybridization, and ligation cycles. In each cycle, the molecule is processed by the restriction enzyme to yield a double-stranded molecule encoding the actual state and the next input symbol. This molecule hybridizes and ligates with the corresponding transition molecule to yield the next molecule to be processed. The processing terminates either when a terminator is reached or no transition rule is applicable. The final molecule anneals to

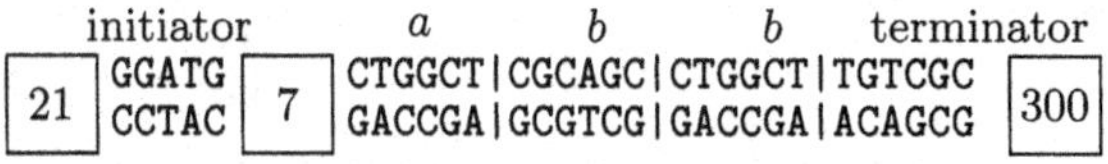

initiator
a
b
b
terminator
21
GGATG CCTAC
7
CTGGCT | CGCAGC | CTGGCT | TGTCGC
GACCGA | GCGTCG | GACCGA | ACAGCG
300

Fig. 6.16 Encoding of input string *aba*.

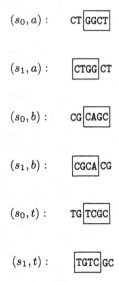

$(s_0, a) :$ CT GGCT

$(s_1, a) :$ CTGG CT

$(s_0, b) :$ CG CAGC

$(s_1, b) :$ CGCA CG

$(s_0, t) :$ TG TCGC

$(s_1, t) :$ TGTC GC

Fig. 6.17 Encoding of symbols and states (t stands for terminator).

$s_0 - D :$ 161 AGCG

$s_1 - D :$ 251 ACAG

Fig. 6.18 Output detection molecules.

an output detector molecule to form an output reporter molecule. This will reveal whether the input string is accepted or not.

Example 6.11. Consider the finite state automaton M given in Figure 6.19, with input alphabet $\Sigma = \{a, b\}$, state set $S = \{s_0, s_1\}$, initial state s_0, and final state set $F = \{s_1\}$. The language of M consists of all strings over Σ

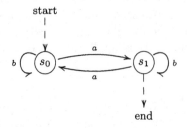

Fig. 6.19 Finite state automaton.

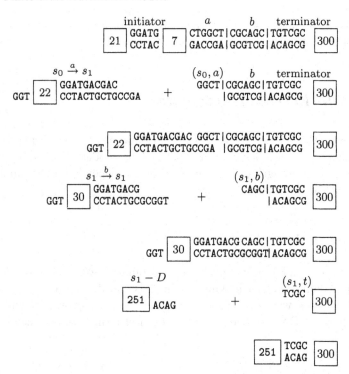

Fig. 6.20 Processing of the input string ab by two restriction, hybridization, and ligation cycles.

with an odd number of a's. The processing of the input string ab is illustrated in Figure 6.20.   ◊

Implementation Issues

This automaton model operates in a completely autonomous fashion so that each molecule induces an independent parallel computation. In the implementation, a computation over 1.5×10^{12} molecules of four symbol inputs (2.5 pmol) rendered output reporter molecules with about 99% yield producing 7.5×10^{11} outputs in $4,000$ s. But each output is the result of five transitions and so the computing performance is of the order of 10^9 transitions per second. As each transition consumes two ATP molecules releasing 1.5×10^{-19} J, the energy comsumption rate is about $1.5 \times 10^{-19} \times 10^9 = 1.5 \times 10^{-10}$ J/s.

The downside is the limited complexity of the two-state two-symbol automaton model (i.e., number of symbols multiplied by number of states). Any increase in complexity is bound from above by the number of different non-palindromic sticky ends. The engineering of a new class of restriction endonucleases with longer spacers or longer sticky ends might allow the implementation of automata with higher complexity.

6.2.2 Length-Encoding Automata

A more general DNA model for finite state automata was recently proposed by J. Kuramochi and Y. Sakakibara (2006). This model is called *length-encoding* because the length of an input string also depends on the number of the states of the automaton.

Data Representation

Let $M = (\Sigma, S, \delta, s_0, F)$ be a finite state automaton with state set $S = \{s_0, \ldots, s_m\}$. Assume that each symbol a in Σ is encoded by a single DNA strand $\sigma(a)$. In this way, each input string $x = a_1 \ldots a_n$ over Σ will be encoded by a single DNA strand consisting of alternating symbol encodings and spacers,

$$\sigma(x) = 5' - X_1 \ldots X_m \sigma(a_1) X_1 \ldots X_m \sigma(a_2) \ldots X_1 \ldots X_m \sigma(a_n) - 3' , \quad (6.5)$$

where $X_1 \ldots X_m$ is the *spacer sequence* (*sp*) consisting of m nucleotides. Moreover, two supplementary sequences are added at both ends of the input molecule, one probe for affinity purification with magnetic beads at the 3' end and primer sequences for PCR at both ends. Thus, the encoded input string has the following form

$$5' - R_1 \ldots R_u X_1 \ldots X_m \sigma(a_1) X_1 \ldots X_m \ldots \sigma(a_n) Y_1 \ldots Y_v Z_1 \ldots Z_w - 3' , \quad (6.6)$$

where $Y_1 \ldots Y_v$ is a probe, and $R_1 \ldots R_u$ and $Z_1 \ldots Z_w$ are primers.

The transition rule $\delta(s_i, a) = s_j$, $a \in \Sigma$, is encoded by the single strand

$$3' - \overline{X}_{i+1} \ldots \overline{X}_m \overline{\sigma(a)} \overline{X}_1 \ldots \overline{X}_j - 5' , \quad (6.7)$$

where $\overline{X}$ denotes the Watson-Crick complement of the nucleotide X, and $\overline{\sigma(a)}$ refers to the Watson-Crick complement of the DNA sequence $\sigma(a)$. Furthermore, a PCR primer is put in front of the input strand, and both a PCR primer and a strand for streptavidin-biotin bonding are placed at the end of the input strand. The weakness of this model is that s_0 must be both initial and final state due to the encoding of the input string.

Example 6.12. Consider the finite state automaton M in Figure 6.23. Put $\sigma(a) = \text{CCGG}$, $\sigma(b) = \text{GCGC}$ and take as spacer the nucleotide A. Then the

Table 6.1 Encoding of state transitions in finite state automaton (Fig. 6.23).

Transition	Encoding
$\delta(s_0, a) = s_1$	$3' - \text{TGGCCT} - 5'$
$\delta(s_0, b) = s_0$	$3' - \text{TCGCG} - 5'$
$\delta(s_1, a) = s_0$	$3' - \text{GGCC} - 5'$
$\delta(s_1, b) = s_1$	$3' - \text{CGCGT} - 5'$

string $x = abab$ is encoded as $5' - \text{ACCGGAGCGCACCGGAGCGC} - 3'$, and the state transitions are encoded as in Table 6.1. ◇

Computations

The molecular automaton operates in three steps: data pre-processing, computation, and output verification. Data pre-processing is accomplished by the following steps:

- Encode the input string by a corresponding single-stranded DNA molecule, and state-transitions and supplementary sequences by short pieces of complementary single-stranded DNA.
- Put all single-stranded DNA molecules into a test tube and let complementary strands hybridize.
- Add ligase into the test tube in order to obtain (partially) double-stranded DNA molecules.

The formation of double-stranded DNA molecules is based on the observation that two consecutive transitions $\delta(s_i, a_l) = s_j$ and $\delta(s_j, a_{l+1}) = s_k$ encode a concatenated strand with a complementary spacer in between:

$$3' - \overline{X}_{i+1} \ldots \overline{X}_m \overline{\sigma(a_l)} \overline{X}_1 \ldots \overline{X}_m \overline{\sigma(a_{l+1})} \overline{X}_1 \ldots \overline{X}_k - 5' . \qquad (6.8)$$

After pre-processing, an accepted input string will correspond to a *completely* double-stranded DNA molecule, while a non-accepted input string will correspond to *partially* double-stranded DNA.

The computation involves the following steps:

- Denaturate the double-stranded DNA molecules into single-stranded DNA.
- Extract the single DNA strands containing biotinylated probe subsequences by streptavidin-biotin bonding with magnetic beads.
- Amplify the extracted single DNA strands with PCR primers.
- Separate the PCR products by length using gel electrophoresis and detect the band corresponding to the length of the input. If this band exists, the input string is accepted; otherwise, it is rejected.

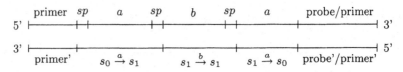

Fig. 6.21 Double-stranded molecule encoding the accepted string *aba*.

Laboratory experiments were carried out for various finite automata with 2 to 6 states for several input strings.

Example 6.13. In view of the previous example, the input string $x = aba$ encoded by the single DNA strand $5' - \text{ACCGGAGCGCACCGG} - 3'$ is accepted by the automaton via the transitions $s_0 \xrightarrow{a} s_1$, $s_1 \xrightarrow{b} s_1$, and $s_1 \xrightarrow{a} s_0$. Annealing and ligation lead to the corresponding double-stranded DNA molecule (Fig. 6.21)

$$5' - \text{ACCGGAGCGCACCGG} - 3'$$
$$3' - \text{TGGCCTCGCGTGGCC} - 5'$$

where primers and probe are not shown. After denaturation, the complementary strands containing probe and primer (probe'/primer') are extracted and amplified, and a band of length equal to the input can be eventually detected.

The input string $x = abb$ encoded by $5' - \text{ACCGGAGCGCAGCGC} - 3'$ is not accepted by the automaton. Annealing and ligation provide a partially double-stranded DNA molecule, because the last transition $s_1 \xrightarrow{b} s_1$ ends with a spacer, while there is no spacer between the last symbol and the probe (Fig. 6.22). After denaturation, the complementary strands consisting of complementary probe and primer (probe'/primer') are extracted and amplified, providing a band of length smaller than the input.   ◇

6.2.3 Sticker Automata

A DNA model for finite state machines was recently proposed by the book authors (2007). It is termed sticker automaton model, because the transition rules are encoded by short single-stranded DNA molecules referred to as stickers. This model generalizes the length-encoding model.

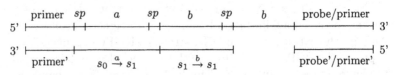

Fig. 6.22 Partially double-stranded molecule encoding the non-accepted string *abb*.

Data Representation

Let $M = (\Sigma, S, \delta, S_0, F)$ be a non-deterministic finite state automaton with state set $S = \{s_0, \ldots, s_{m-1}\}$. Suppose each symbol a in Σ is encoded by a single DNA strand $\sigma(a)$. In this way, each input string $x = a_1 \ldots a_n$ over Σ will be encoded by a single DNA strand consisting of alternating symbol encodings and spacers,

$$\sigma(x) = 5' - I_1 X_0 \ldots X_m \sigma(a_1) \ldots X_0 \ldots X_m \sigma(a_n) X_0 \ldots X_m I_2 - 3' , \quad (6.9)$$

where $X_0, \ldots, X_m$ is the *spacer sequence*, I_1 is the *initiator sequence*, and I_2 is the *terminator sequence*.

The transition rule $\delta(s_i, a) = s_j$, $a \in \Sigma$, is encoded by the single strand

$$3' - \overline{X}_{i+1} \ldots \overline{X}_m \overline{\sigma(a)} \overline{X}_0 \ldots \overline{X}_j - 5' , \quad (6.10)$$

where $\overline{X}$ denotes the Watson-Crick complement of the nucleotide X, and $\overline{\sigma(a)}$ refers to the Watson-Crick complement of the DNA sequence $\sigma(a)$. Each initial state $s_i \in S_0$ is encoded by the single strand

$$3' - \overline{I}_1 \overline{X}_0 \ldots \overline{X}_i - 5' \quad (6.11)$$

and each final state s_j is encoded by the single strand

$$3' - \overline{X}_{j+1} \ldots \overline{X}_m \overline{I}_2 - 5' . \quad (6.12)$$

The hardware is composed of a single enzyme, Mung Bean nuclease or S1 nuclease, that will be explained later. In contrast to the Shapiro model, the software does not contain any recognition site for restriction enzymes, and symbols and states are separately encoded.

Example 6.14. Consider the finite state automaton M in Figure 6.23, with state set $S = \{s_0, s_1\}$, input alphabet $\Sigma = \{a, b\}$, initial state s_0, and final state set $F = \{s_0\}$. In view of this automaton, the schematic encoding of the input string aba (Fig. 6.24), the encoding of initiator and

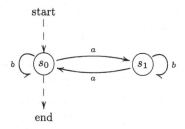

Fig. 6.23 Finite state automaton.

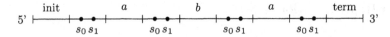

Fig. 6.24 Encoding of input data string *aba*.

terminator (Fig. 6.25), and the encoding of transition rules (Fig. 6.26) are illustrated.

Put $\sigma(a) = \text{CCGG}$ and $\sigma(b) = \text{GCGC}$. Take as spacers the dinucleotide sequences $X_0 = \text{AA}$, $X_1 = \text{GA}$, and $X_2 = \text{AG}$. Let the initiator sequence I_1 and terminator sequence I_2 be GG and GC, respectively. The string $x = aba$ is then encoded as

$$5' - \text{GGAAGAAGCCGGAAGAAGGCGCAAGAAGCCGGAAGAAGGC} - 3' \,.$$

The initial state is given as $3' - \text{CCTT} - 5'$, the final state is referred to as $3' - \text{CTTCCG} - 5'$, and the state transitions are encoded as in Table 6.2.   ◇

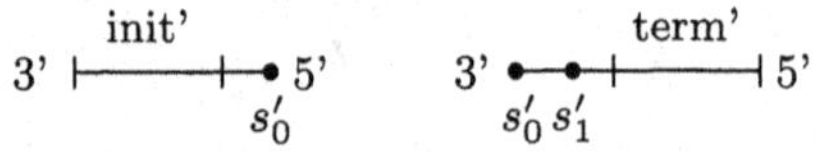

$$
\begin{array}{cc}
\text{init'} & \text{term'}\\
3' \vdash\!\!\!-\!\!\!-\!\!\!\bullet\, 5' & 3' \bullet\!\!-\!\!\bullet\!\!+\!\!\!-\!\!\!-\!\!\dashv 5'\\
\quad s_0' & s_0'\, s_1'
\end{array}
$$

Fig. 6.25 Encoding of initial and final state s_0.

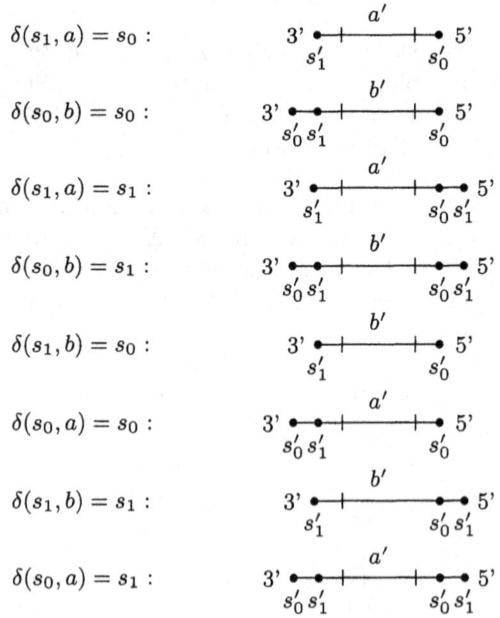

$$
\begin{array}{ll}
\delta(s_1, a) = s_0: & \quad 3' \bullet\!\!+\!\!\!\xrightarrow{\;\;a'\;\;}\!\!+\!\bullet\, 5'\\
& \qquad s_1' \qquad\qquad s_0'\\[4pt]
\delta(s_0, b) = s_0: & \quad 3' \bullet\!\!-\!\!\bullet\!\!+\!\!\!\xrightarrow{\;\;b'\;\;}\!\!+\!\bullet\, 5'\\
& \qquad s_0'\, s_1' \qquad\qquad s_0'\\[4pt]
\delta(s_1, a) = s_1: & \quad 3' \bullet\!\!+\!\!\!\xrightarrow{\;\;a'\;\;}\!\!+\!\bullet\!\!-\!\!\bullet\, 5'\\
& \qquad s_1' \qquad\qquad s_0'\, s_1'\\[4pt]
\delta(s_0, b) = s_1: & \quad 3' \bullet\!\!-\!\!\bullet\!\!+\!\!\!\xrightarrow{\;\;b'\;\;}\!\!+\!\bullet\!\!-\!\!\bullet\, 5'\\
& \qquad s_0'\, s_1' \qquad\qquad s_0'\, s_1'\\[4pt]
\delta(s_1, b) = s_0: & \quad 3' \bullet\!\!+\!\!\!\xrightarrow{\;\;b'\;\;}\!\!+\!\bullet\, 5'\\
& \qquad s_1' \qquad\qquad s_0'\\[4pt]
\delta(s_0, a) = s_0: & \quad 3' \bullet\!\!-\!\!\bullet\!\!+\!\!\!\xrightarrow{\;\;a'\;\;}\!\!+\!\bullet\, 5'\\
& \qquad s_0'\, s_1' \qquad\qquad s_0'\\[4pt]
\delta(s_1, b) = s_1: & \quad 3' \bullet\!\!+\!\!\!\xrightarrow{\;\;b'\;\;}\!\!-\!\bullet\!\!-\!\!\bullet\, 5'\\
& \qquad s_1' \qquad\qquad s_0'\, s_1'\\[4pt]
\delta(s_0, a) = s_1: & \quad 3' \bullet\!\!-\!\!\bullet\!\!+\!\!\!\xrightarrow{\;\;a'\;\;}\!\!+\!\bullet\!\!-\!\!\bullet\, 5'\\
& \qquad s_0'\, s_1' \qquad\qquad s_0'\, s_1'
\end{array}
$$

Fig. 6.26 Encoding of all 8 transitions in a two-symbol two-state automaton.

Table 6.2 Encoding of state transtitions in finite state automaton (Fig. 6.23).

Transition	Encoding
$\delta(s_0, a) = s_1$	$3' - \text{CTTCGGCCTTCT} - 5'$
$\delta(s_0, b) = s_0$	$3' - \text{CTTCCGCGTT} - 5'$
$\delta(s_1, a) = s_0$	$3' - \text{TCGGCCTT} - 5'$
$\delta(s_1, b) = s_1$	$3' - \text{TCCGCGTTCT} - 5'$

Computation

The molecular automaton operates in three steps: data pre-processing, computation, and output verification. Data pre-processing is accomplished by the following steps:

- Encode the input string into the corresponding single-stranded DNA molecule, and state-transitions and supplementary sequences into short pieces of complementary single-stranded DNA.
- Put all single-stranded DNA molecules into a test tube and let complementary strands hybridize.
- Add ligase into the test tube in order to obtain (partially) double-stranded DNA molecules.

The formation of double-stranded DNA molecules is based on the observation that two consecutive transitions $\delta(s_i, a_l) = s_j$ and $\delta(s_j, a_{l+1}) = s_k$ encode a concatenated strand with a complementary spacer inbetween:

$$3' - \overline{X}_{i+1} \ldots \overline{X}_m \overline{\sigma(a_l)} \overline{X}_0 \ldots \overline{X}_m \overline{\sigma(a_{l+1})} \overline{X}_0 \ldots \overline{X}_k - 5' \,. \tag{6.13}$$

After pre-processing, an accepted input string will correspond to a complete, double-stranded DNA molecule, while a non-accepted input string will correspond to partially double-stranded DNA.

The computation is carried out by Mung Bean nuclease or S1 nuclease. Mung Bean nuclease and S1 nuclease remove single nucleotides in a stepwise manner from a single-stranded DNA molecule. S1 nuclease can also occasionally introduce single-stranded breaks into a double-stranded DNA. Therefore, Mung Bean nuclease is preferable to S1 nuclease for most applications because it has lower intrinsic activity on duplex DNA. Both enzymes are able to degrade the single-stranded region in a non-accepted input string. As a consequence, complete double-stranded DNA molecules corresponding to accepted input strings will remain intact after enzyme digestion.

Finally, gel electrophoresis can be employed to separate DNA molecules by size so that accepted input strings can be detected. However, this would require knowledge of the lengths of the molecules in advance. On the other hand, the DNA molecule encoding an accepted input string has both an initiator and terminator sequence. Therefore, PCR may be used to detect molecules corresponding to accepted input strings.

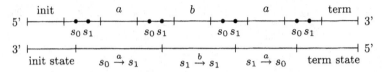

Fig. 6.27 Double-stranded molecule encoding accepted string *aba*.

Example 6.15. In view of the previous example, the input string $x = aba$ is accepted by the automaton M via the transitions $s_0 \overset{a}{\to} s_1$, $s_1 \overset{b}{\to} s_1$, and $s_1 \overset{a}{\to} s_0$. Annealing and ligation lead to a complete, double-stranded DNA molecule (Fig. 6.27).

The input string $x = abb$ is not accepted by the automaton. Annealing and ligation provide a partially double-stranded DNA molecule, which can have one of three forms (Figs. 6.28–6.30).   ◇

Behavioral Simulation

Consider the sticker automaton model for the finite state machine that accepts all strings with an even number of a's over the alphabet $\{a, b\}$ (Fig. 6.23). Now this model is implemented by a chemical reacting system and the time evolution of this system is analyzed by a stochastic simulation

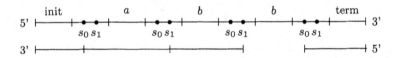

Fig. 6.28 Non-accepted input string *abb*: non-existing transition rule $\delta(s_1, b) = s_0$.

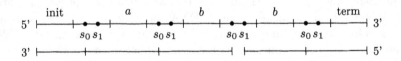

Fig. 6.29 Non-accepted input string *abb*: illegal transition rule $\delta(s_0, b) = s_0$.

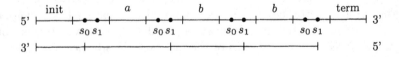

Fig. 6.30 Non-accepted input string *aba*: non-accepted final state.

method, Gillespie's First Reaction Method. To this end, we conduct simulations for the accepted input aba as positive control and the rejected input abb as negative control.

Subsequently, we denote (partially) double-stranded molecules as words in which the input molecules are used as building blocks. The input molecules for aba and abb are represented as $I\_\_a\_\_b\_\_a\_\_I$ and $I\_\_a\_\_b\_\_b\_\_I$, respectively. In view of the input abb, a partially double-stranded molecule consisting of input molecule and annealed initial state molecule is given by $I_1\_a\_\_b\_\_b\_\_I$. Similarly, a partially double-stranded molecule consisting of input molecule and annealed final state molecule is denoted as $I\_\_a\_\_b\_\_b\_2I$. Moreover, a partially double-stranded molecule consisting of input molecule $I\_\_a\_\_b\_\_b\_\_I$ and annealed transition rule $s_0 \xrightarrow{a} s_1$ is represented as $I\_0 a_1\_b\_\_b\_\_I$. Using this notation, a feasible chain of hybridization reactions for the input aba is the following:

$$I\_\_a\_\_b\_\_a\_\_I + I_1 \rightleftharpoons I_1\_a\_\_b\_\_a\_\_I$$

$$I_1\_a\_\_b\_\_a\_\_I + s_0 \xrightarrow{a} s_1 \rightleftharpoons I_{10}a_1\_b\_\_a\_\_I$$

$$I_{10}a_1\_b\_\_a\_\_I + s_1 \xrightarrow{b} s_1 \rightleftharpoons I_{10}a_{11}b_1\_a\_\_I$$

$$I_{10}a_{11}b_1\_a\_\_I + s_1 \xrightarrow{a} s_0 \rightleftharpoons I_{10}a_{11}b_{11}a_0\_I$$

$$I_{10}a_{11}b_{11}a_0\_I + {}_2I \rightleftharpoons I_{10}a_{11}b_{11}a_{02}I \ .$$

The final product $I_{10}a_{11}b_{11}a_{02}I$ is a double-stranded DNA molecule encoding the accepted input aba. Other potential products that do not form complete double-stranded DNA molecules are assumed to be degraded by a nuclease.

In view of the simulations, the symbols and spacers are encoded by oligonucleotides of length of 10 nt. Initial populations of 10,000 molecules per input molecule and sticker molecule are mixed in a volume of 10^{-15} litres, which is held at temperature $T = 40°C$. Here perfect and complete matching between sticker molecules and input molecules is assumed, while products with partial or overlapping hybridizations are eliminated.

In terms of the accepted input aba, there are more than 1,800 reactions (including reverse reactions) with a total of 74 distinct reactants (Fig. 6.31). During the simulation, the concentration of the molecules for the transition $s_0 \xrightarrow{b} s_0$ drastically reduces during the first minute of reaction and then remains constant over time, because this transition is not required for the assembly of the accepted input. In fact, the concentration of the molecules for the transition $s_0 \xrightarrow{b} s_0$ is always higher than that of the accepted duplex over time. Competitive reactions during simulations provoked some illegal transition molecules to anneal with input strands forming incomplete double-stranded molecules, lowering the efficiency of the computation. This effect can be counteracted by increasing the concentrations of the participating strands.

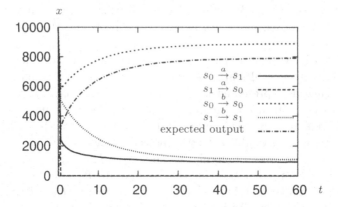

Fig. 6.31 Simulation results of accepted input *aba* with an initial population of 10,000 molecules per input and sticker molecule (time (min) vs. number of molecules).

In view of the non-accepted input *abb*, there are more than 1,170 reactions (including reverse reactions) with a total of 67 different reactants (Fig. 6.32). During the simulation, the most stable molecule turned out to be $I_{11}a_{00}b_{00}b_{02}I$. The average number of these molecules after 1 h was about 5,000 out of 10,000. These molecules consumed almost all molecules for the transition $s_0 \overset{b}{\to} s_0$, while the molecules for the transitions $s_1 \overset{b}{\to} s_1$ and $s_0 \overset{a}{\to} s_1$ were hardly required. None of the duplexes generated during the simulation resulted in complete double-stranded DNA molecules.

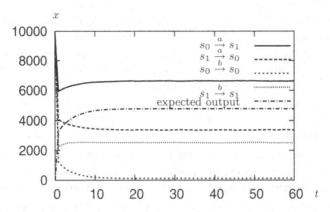

Fig. 6.32 Simulation results of non-accepted input *abb* with an initial population of 10,000 molecules per input and sticker molecule (time (min) vs. number of molecules).

6.2.4 Stochastic Automata

Consider a DNA model for a deterministic or non-deterministic finite state machine, which contains DNA molecules that individually encode initial states, final states, and transition rules. The relative molar concentrations of the molecules encoding initial and final states can be viewed as probabilities of the initial and final states, respectively. Moreover, the relative molar concentrations of the molecules encoding transition rules can be considered as conditional probabilities of the transition rules. In this way, the DNA model implements a stochastic variant of the finite state machine as first proposed by E. Shapiro and coworkers (2004). The automaton models described in this section fulfill the above requirements and hence can also implement stochastic parsers.

6.3 DNA Hairpin Model

The hairpin model was invented by M. Hagiya and coworkers in 1998 to evaluate Boolean functions. This model was further developed by E. Winfree (1998) and termed *polymerization stop* since it extends the 3' end of hairpin loops in single-stranded DNA molecules by stopped polymerization. Later, the hairpin model was modified so that the hairpin loops in single-stranded DNA can be enzymatically digested. This model allows famous NP-complete problems by autonomous DNA computations to be tackled.

6.3.1 Whiplash PCR

Today, polymerization stop is known under the notion of *whiplash PCR*, so dubbed by L. Adleman.

One-Shot Boolean Expressions

A *one-shot Boolean expression* or *μ-formula* is a Boolean expression that is inductively defined as follows:

- Each Boolean variable is a μ-formula.
- If f is a μ-formula then its negation $\overline{f}$ is a μ-formula.
- If f and g are μ-formulas, then so are their conjunction fg and disjunction $f + g$, provided that f and g do not share a common variable.

In view of the last definition, μ-formulas contain each variable at most once. For instance, μ-formulas are $(x_1 + \overline{x}_2)(\overline{x}_3 + x_4)$ and $x_1\overline{x}_2 + x_3 + x_4$.

These formulas can be evaluated by particularly simple binary decision diagrams. Such a diagram is given by a labelled acyclic digraph with three distinguished vertices, b (of indegree 0), e_0 and e_1 (both of outdegree 0), while the remaining vertices correspond one-to-one with the variables. A labelled edge $x_i \xrightarrow{X} x_j$, $X \in \{0,1\}$, indicates that the variable x_i is assigned the truth value X and x_j is evaluated next. The diagram provides for each truth assignment of the variables a path from the vertex b to one of the vertices e_0 or e_1 so that the μ-formula evaluates to X if and only if the path reaches the vertex e_X (Fig. 6.33).

Data Representation

Let f be a μ-formula given by a binary decision diagram, and suppose there is an assignment of truth values to the variables of the μ-formula. The aim is to evaluate the μ-formula f according to the truth assignment by an autonomous molecular computation. To this end, truth assignment (data part), μ-formula (program part), and current state (head) are encoded by one single-stranded DNA molecule. The data part encodes the truth assignment: Each assignment $x_i = X$, $X \in \{0,1\}$, is encoded by the pair of oligonucleotides (x_i^X, x_i). The program part encodes the corresponding binary decision diagram: Each edge $x_i \xrightarrow{X} x_j$, $X = 0, 1$, is encoded by the pair of oligonucleotides (x_j, x_i^X), and

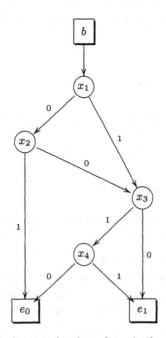

Fig. 6.33 Binary decision diagram for the μ-formula $f = (x_1 + \overline{x}_2)(\overline{x}_3 + x_4)$.

Fig. 6.34 Encoding of truth assignment (data), μ-formula (program), and initial state $b \to x_1$.

each edge $x_i \xrightarrow{X} e_Y$, $X, Y \in \{0, 1\}$, is encoded by the pair of oligonucleotides (e_Y, x_i^X). Finally, the edge $b \to x_i$ corresponds to the initial state and is encoded by the Watson-Crick complement $\overline{x}_i$ of the oligonucleotide encoding the variable x_i.

Example 6.16. Consider the μ-formula $f = (x_1 + \overline{x}_2)(\overline{x}_3 + x_4)$ and the truth assignment $x_1 = 1$, $x_2 = 0$, $x_3 = 1$, and $x_4 = 0$. The corresponding single-stranded DNA molecule (Fig. 6.34) has the data part

$$(x_4^0, x_4)(x_3^1, x_3)(x_2^0, x_2)(x_1^1, x_1) \, ,$$

the program part

$$(e_0, x_2^1)(x_3, x_2^0)(e_1, x_4^1)(e_0, x_4^0)(x_4, x_3^1)(e_1, x_3^0)(x_3, x_1^1)(x_2, x_1^0) \, ,$$

and the initial state $\overline{x}_1$. $\Diamond$

Moreover, consecutive pairs of oligonucleotides in the data and program parts are separated by a so-called *stopper sequence* consisting of nucleotides whose complement is missing in the polymerization buffer. These sequences allow termination of polymerization. For instance, the triplet GGG can be used as a stopper sequence if cytosines are missing in the polymerization buffer.

Evaluation of One-Shot Boolean Expressions

A μ-formula is evaluated according to a given truth assignment by state transitions implemented via molecular operations. In each transition, the head reads information alternatively from the data and program part and changes its state. For this, the single-stranded DNA molecule forms an intramolecular hairpin structure between the current state and a substrand in the data or program part. Then the 3' end of the current state is extended by one symbol to obtain the next state using polymerization stop.

The transitions are realized by thermal cycles. If the current state is the Watson-Crick complement of a variable x_i, then a hairpin structure is formed with the encoded variable in the data part, (x_i^0, x_i) or (x_i^1, x_i). Polymerization extends the molecule with the Watson-Crick complement of the assignment x_i^0 or x_i^1, and the hairpin structure is denatured (Fig. 6.35). If the current state is the Watson-Crick complement of an assignment x_i^0 or x_i^1, then a hairpin structure is formed with the encoded variable in the program part, (x_j, x_i^0) or (x_j, x_i^1). Polymerization extends the molecule with the

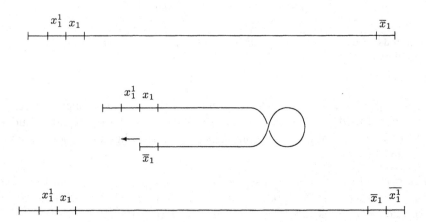

Fig. 6.35 A thermal cycle in the computation for variable evaluation: The variable x_1 is assigned the value 1 by hairpin formation and polymerization stop.

Watson-Crick complement of the variable x_j, and the hairpin structure is denaturated (Fig. 6.36). Thus, two consecutive thermal cycles are necessary to assign the truth value to the actual variable in the data part, and to pass to the next variable in the program part. Hence, the computation evolves in $2m$ thermal cycles, where m is the length of the path in the binary decision diagram corresponding to the truth assignment.

Example 6.17. In view of the μ-formula $f = (x_1 + \overline{x}_2)(\overline{x}_3 + x_4)$ and the truth assignment $x_1 = 1$, $x_2 = 0$, $x_3 = 1$, and $x_4 = 0$, the traversed path in the

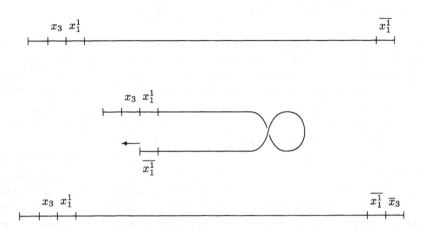

Fig. 6.36 A thermal cycle in the computation for state transition: The transition $x_1 \xrightarrow{1} x_3$ in the binary decision diagram is performed by hairpin formation and polymerization stop.

associated binary decision diagram (Fig. 6.33) is $b \to x_1 \xrightarrow{1} x_3 \xrightarrow{1} x_4 \xrightarrow{0} e_0$. This path corresponds to the following sequence of states extending the 3' end of the molecule: $\overline{x}_1$ (initial state), x_1^1, $\overline{x}_3$, x_3^1, $\overline{x}_4$, x_4^0, and $\overline{e}_0$. $\diamond$

Implementation Issues

The evaluation of μ-formulas requires that intramolecular reactions will exclusively occur. For this, the experimental conditions need to be properly adjusted so that intramolecular reactions are facilitated, while intermolecular reactions are non-favorable. This can be achieved by two simple methods: keeping the concentration of DNA molecules sufficiently low and quickly cooling down DNA molecules during annealing.

In view of consecutive thermal cycles, it is highly probable that the previous hairpin structure is formed again and therefore no successive transitions occur. To avoid such situations, the probability of the latest hairpin structure must be reasonably large compared with the previous ones. This can be achieved by designing the oligonucleotides so that the latest hairpin structure has higher stability than the previous ones. Moreover, hairpin formation may be facilitated by including a spacer sequence between the program part and the initial state. Otherwise, the hairpin loop might be too short.

Finally, artificial nucleosides such as iso-cytosine may be used for stopper sequences. In this way, all four nucleotides can be employed for encoding the data and program part, and the GC-contents will be easier to adjust to an appropriate level.

6.3.2 Satisfiability

The whiplash PCR model was slightly modified by M. Hagiya and coworkers (2000) so that the satisfiability problem can be tackled in an autonomous manner. To see this, let F be a Boolean expression in n variables given in CNF such as

$$F = (x_2 + \overline{x}_3)(x_1 + \overline{x}_2 + \overline{x}_3)(\overline{x}_1 + x_2) .$$

Define a *literal string* of the Boolean expression F as a conjunction of literals in F so that one literal per disjunctive clause is selected. For instance, literal strings of the expression F are

$$x_2 x_1 \overline{x}_1, \quad \overline{x}_3 \overline{x}_2 \overline{x}_1, \quad \text{and} \quad \overline{x}_3 x_1 x_2 .$$

Theorem 6.18. *A Boolean expression F in CNF is satisfiable if and only if there is a literal string of F that does not involve a variable and its negation.*

Proof. Let $F = c_1 \ldots c_m$ be a Boolean expression in CNF, where $c_1, \ldots, c_m$ are (disjunctive) clauses over the variables $x_1, \ldots, x_n$. Let $L = z_{i_1} \ldots z_{i_m}$ be a literal string of F so that z_{i_j} is a literal occurring in clause c_j and L does not involve a variable and its negation, $1 \le j \le m$. If z_{i_j} is a variable, assign to the variable the value 1. Otherwise, assign to the variable the value 0. In view of the hypothesis on L, this assignment is well-defined. It follows that each clause c_i is satisfied, $1 \le i \le m$, and hence F is satisfied.

Conversely, if each literal string of F involves at least one variable and its negation, then the expression F is not satisfiable.   □

For instance, the literal string $\bar{x}_3 x_1 x_2$ in F fulfills the hypothesis in Theorem 6.18. Thus, in view of the assignment $x_1 = 1$, $x_2 = 1$, and $x_3 = 0$, the Boolean expression F is satisfiable.

Data Representation

In view of the DNA hairpin model of the satisfiability problem, each literal in the ith clause of the Boolean expression F is encoded by a double-stranded DNA molecule with sticky ends. These sticky ends provide links to the $i-1$th clause to the left and the ith clause to the right (Fig. 6.37). In terms of this encoding, double-stranded DNA molecules can be formed by annealing and ligation of literals, which correspond to the literal strings of F (Fig. 6.38).

Each variable and its negation are encoded by short Watson-Crick complementary DNA strands so that the center contains the sequence

$$5' - \text{CCAN}_1\text{N}_2\text{N}_3\text{N}_4\text{N}_5 | \text{N}_6\text{TGG} - 3'$$
$$3' - \text{GGT}\bar{\text{N}}_1 | \bar{\text{N}}_2\bar{\text{N}}_3\bar{\text{N}}_4\bar{\text{N}}_5\bar{\text{N}}_6\text{ACC} - 5'$$

where N_i is any nucleotide and $\bar{\text{N}}_i$ denotes its Watson-Crick complement. This sequence hosts the recognition site of the restriction endonuclease BstXI, where the cutting sites are indicated by vertical bars.

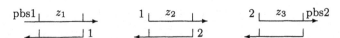

Fig. 6.37 Encoding of literals: A literal z_i from the ith clause is encoded by a double-stranded DNA molecule that contains linker numbers at both sides. Literals from the first and last clause contain primer binding sites pbs1 and pbs2 as prefix and postfix, respectively. Arrows indicate 5' to 3' direction.

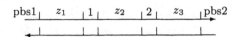

Fig. 6.38 Self-assembled literal string from literals.

Algorithm Description

The autonomous computation starts by designing an initial test tube consisting of a multiset of double-stranded DNA encoding the literals. The literals self-assemble by annealing and ligation to yield literal strings. These literal string DNA molecules are denatured at low concentration. The resulting single-stranded DNA molecules are allowed to refold in an intramolecular manner. This means that literal strings containing a variable and its negation eventually form hairpins, thanks to the encoding of literals (Fig. 6.39). This computation is performed solely by controlling the temperature.

The hairpins formed in single-stranded DNA molecules can be digested by the restriction endonuclease BstXI. For this, notice that the enzyme BstXI has an optimal incubation temperature of 55°C. By Theorem 6.18, the non-digested single-stranded molecules of length km nt correspond to literal strings that satisfy the Boolean expression, where k is the length of each literal strand in nt and m is the number of clauses in the Boolean expression. Thus, length separation of the DNA molecules using gel electrophoresis eventually provides literal strands of length km nt. These strands yield a solution of the satisfiability problem for the Boolean expression.

Implementation Issues

Hagiya and coworkers demonstrated the feasibility of this algorithm for a Boolean expression with six variables and ten clauses, where each variable was encoded by a strand of 30 bp. A major drawback of the described autonomous DNA algorithm for solving the satisfiability problem lies in the required amount of DNA. While common DNA filtering algorithms require 2^n molecules for the encoding of the truth value assignments of n variables, the hairpin model generates 3^m strings for m clauses for the 3-SAT problem, where m can be a multiple of n for harder instances.

6.3.3 Hamiltonian Paths

The hairpin model was used by the authors of the book (2005) to implement an autonomous version of Adleman's first experiment. However, the approach

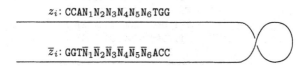

z_i: CCAN$_1$N$_2$N$_3$N$_4$N$_5$N$_6$TGG

$\bar{z}_i$: GGTN̄$_1$N̄$_2$N̄$_3$N̄$_4$N̄$_5$N̄$_6$ACC

Fig. 6.39 Single-stranded DNA literal string with a formed hairpin.

is different from that employed to solve the satisfiability problem: DNA hairpin formation is driven by self-annealing of palindromic sequences, which are rather uncommon in DNA computing models.

Let G be a directed graph and let v_s be a vertex in G. The objective is to find Hamiltonian paths in G that begin with v_s as initial vertex. The following DNA hairpin algorithm solves the Hamiltonian path problem for the graph G in six steps:

1. Dephosphorylate the single-stranded DNA molecules encoding the initial vertex.
2. Generate random paths in the graph.
3. Keep only those paths that begin with the initial vertex.
4. Keep only those paths that contain each vertex at most once.
5. Keep only those paths that have n vertices, where n is the number of vertices of G.
6. Read out Hamiltonian paths (if any).

Data Representation

The information given by the graph G is encoded as in Adleman's model (Sect. 5.1.1). However, the encoding of each vertex contains a *palindromic region* flanked on each side by thymine nucleotides (Table 6.3). This encoding has the general form

$$5' - T_1 T_2 \ldots T_k\ N_1 N_2 \ldots N_l\ \text{GGCC}\ \bar{N}_l \ldots \bar{N}_2 \bar{N}_1\ T_1 T_2 \ldots T_k - 3'\ , \qquad (6.14)$$

where T_i stands for thymine, N_j stands for any nucleotide, and $\bar{N}_j$ stands for the Watson-Crick complement of N_j. The middle part of the palindromic region contains the recognition site of the restriction endonuclease HaeIII,

$$5' - \text{GG} \mid \text{CC} - 3'$$
$$3' - \text{CC} \mid \text{GG} - 5'$$

Table 6.3 Encoding of the vertices in Adleman's graph.

Vertex	Encoding (22 bp)
v_0	$5' - $ TTTAGCAGTGGCCACTGCTTTT $ - 3'$
v_1	$5' - $ TTTTTGTAGGGCCCTACAATTT $ - 3'$
v_2	$5' - $ TTTTCCATCGGCCGATGGATTT $ - 3'$
v_3	$5' - $ TTTCAGTCAGGCCTGACTGTTT $ - 3'$
v_4	$5' - $ TTTTCAGCTGGCCAGCTGATTT $ - 3'$
v_5	$5' - $ TTTCGACTGGGCCCAGTCGTTT $ - 3'$
v_6	$5' - $ TTTTCTGACGGCCGTCAGATTT $ - 3'$

where the cutting sites are indicated by vertical bars. The purpose of the thymine prefix and suffix is to minimize the self-annealing of the DNA strands encoding the vertices.

Algorithm Description

In the first step of the algorithm, the single-stranded DNA molecules encoding the initial vertex are dephosphorylated with alkaline phosphatase, which catalyzes the release of 5'-phosphate groups from DNA. Dephosphorylating these strands will guarantee that they will not serve as targets for lambda exonuclease, which only recognizes 5'-phosphorylated termini.

In the second step, a combinatorial path library is generated by annealing and ligation of all DNA strands encoding the vertices and edges as in Adleman's approach.

The third step employs lambda exonuclease. This enzyme acts in the 5' to 3' direction, catalyzing the removal of strands with 5'-phosphate group. Thus, lambda exonuclease digests all strands of edges and strands of vertices that do not begin with the initial vertex dephosphorylated in the first step. The remaining non-digested molecules are single DNA strands that correspond to paths beginning with the initial vertex (Fig. 6.40). These single-stranded DNA molecules eventually form hairpins if they correspond to paths with two or more repeated vertices. The partially double-stranded region of such a hairpin structure contains the HaeIII recognition site.

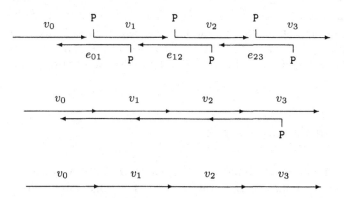

Fig. 6.40 Generation of initial test tube: First, random paths in the graph are formed via annealing, here (v_0, v_1, v_2, v_3). The 5'-phosphate group of the single-stranded DNA molecules associated with the initial vertex v_0 was removed beforehand by alkaline phosphatase. P stands for a 5'-phosphate group. Second, ligase bonds together single-stranded DNA molecules to provide double-stranded molecules that encode paths in the graph. Third, lambda exonuclease selectively digests all 5'-phosphorylated DNA strands, leaving only those single-stranded DNA molecules that begin with the initial vertex.

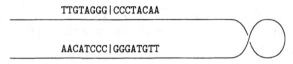

TTGTAGGG | CCCTACAA

AACATCCC | GGGATGTT

Fig. 6.41 Single-stranded DNA corresponding to a path with two occurrences of vertex v_1 (Table 6.3) and formed hairpin.

In the fourth step, the restriction endonuclease HaeIII is added to the solution that cuts within the partially double-stranded region of all DNA strands that have formed hairpins (Fig. 6.41). The remaining non-digested single-stranded DNA will correspond to paths in the graph which contain each vertex at most once. Here an issue that needs to be addressed further is the optimal length of the thymine pre- and suffixes. It depends on the length of the palindromic region and should be designed so that self-annealing of vertex strands can be avoided. Moreover, the competition between intra- and interstrand annealing can be reduced by designing vertex strands with longer palindromic regions that do not separate when two different single strands collide.

In the fifth step, the remaining molecules are separated by size via gel electrophoresis. To this end, notice that there are no molecules longer than $(7 \times 22 =)$ 154 nt, while most DNA fragments of the reaction stay at the bottom of 22 nt. The 154 nt fragment indicates that there is a solution of the seven-vertex Hamiltonian path problem. Finally, the 154 nt fragment needs to be analyzed to provide the actual order of the vertices in the solution molecules, as in Adleman's experiment.

The above procedure was successfully implemented in the laboratory for Adleman's graph with initial vertex v_0 (Table 6.4). It appears that this method can be scaled up and automated to solve an instance of the Hamiltonian path problem with 20 cities. This estimate is in accordance with the largest instance (20 variables) of the SAT problem solved by DNA computing (Sect. 5.2.4).

6.3.4 Maximum Cliques

The maximum clique problem can be solved by a DNA hairping algorithm that is quite similiar to the previous one. This algorithm proposed by the authors (2005) uses the same hardware (enzymes) as that for the Hamiltonian path problem, but the encoding of the graph considered is different.

Let G be a graph. The maximal clique problem for G can be solved by the following DNA hairpin algorithm:

1. Generate random subgraphs of G.
2. Keep only those subgraphs that provide cliques in G.
3. Detect maximal cliques in G.

Table 6.4 Experimental protocol for DNA hairpin experiment of Hamiltonian paths. After steps 2 and 4, DNA samples were purified via nucleotide purification kit and soluted in 50 µl extraction buffer, and 8% polyacrylamide gel was run at 70 V.

Step 1: Dephosphorylation of Initial Vertex	
v_0 strand (200 µM)	10 µl
alkaline phosphatase buffer 10×	2 µl
alkaline phosphatase	2 µl
water	6 µl
Incubate mixture at 37°C for 3 h	
Denaturate enzyme at 85°C for 16 min.	
Step 2: Linking	
v_0 strand (100 µM)	5 µl
every other vertex strand (200 µM)	2.5 µl
each edge strand (400 µM)	2.5 µl
ligase reaction buffer 10×	6 µl
water	8 µl
T4 DNA ligase	5 µl
Incubate mixture at 22°C for 3 h	
denaturate enzyme at 85°C for 10 min.	
Step 3: Exonuclease Digestion	
DNA solution (from step 2)	30 µl
reaction buffer 10×	4 µl
water	4 µl
lambda exonuclease	2 µl
Incubate mixture at 37°C for 45 min	
denaturate enzyme at 80°C for 20 min.	
Step 4: Hairpin Digestion	
DNA solution (from step 3)	40 µl
reaction buffer 10×	6 µl
BSA 100×	0.6 µl
water	10.4 µl
HaeIII	3 µl
Incubate mixture at 37°C for 45 min	
cool down to 4°C to stop reaction.	

In order to apply the hairpin model, we pass from the graph G to its complementary graph G' (Fig. 6.42), and observe that the cliques in G are in one-to-one correspondence with the independent sets in G'. In particular, the maximum cliques in G correspond to the maximal independent sets in G'. In view of this correspondence, the above algorithm also solves the maximum independent set problem for G':

1. Generate random subgraphs of G'.
2. Keep only those subgraphs that provide independent sets in G'.
3. Detect maximal independent sets in G'.

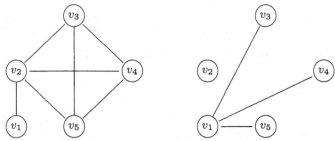

Fig. 6.42 A graph G and its complement G'.

Data Representation

The information given by the graph G' is represented by an *adjacency list*, which provides for each vertex the sequence of incident edges.

Example 6.19. The graph G in Figure 6.42 has the adjacency list given in Table 6.5.   ◇

The adjacency list of G' is encoded by single-stranded DNA molecules called *vertex templates*. Each vertex template corresponds to a vertex v in G' and consists of initial linker l_1, encoding of v, encodings of all edges incident with v, and final linker l_2 (Fig. 6.43). Moreover, all vertices and edges in G' are encoded in palindromic form, containing in the center the recognition site of the restriction endonuclease HaeIII,

$$5' - \mathrm{GG|CC} - 3'$$
$$3' - \mathrm{CC|GG} - 5'$$

where the cutting sites are indicated by vertical bars (Table 6.6). Furthermore, all vertex templates are of the same length. For this, there is a constant non-palindromic encoding sequence called *spacer* (sp) used as a filler. Finally, for each vertex template, part of its concentration is treated with alkaline phosphatase, which dephosphorylates these strands by removing the 5'-phosphate group. Notice that a dephosphorylated vertex template cannot be

Table 6.5 Adjacency list of graph G (Fig. 6.42).

Vertex	List of Edges
v_1	(v_1v_2)
v_2	$(v_1v_2, v_2v_3, v_2v_4, v_2v_5)$
v_3	(v_2v_3, v_3v_4, v_3v_5)
v_4	(v_2v_4, v_3v_4, v_4v_5)
v_5	(v_2v_5, v_3v_5, v_4v_5)

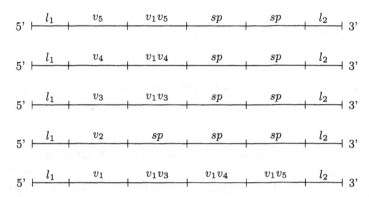

Fig. 6.43 Vertex templates of graph G'.

a substrate for lambda exonuclease, which only recognizes 5'-phosphorylated termini.

Algorithm Description

The first step of the algorithm forms subsets of vertices in G' and thus subgraphs of G' by linking vertex templates. To this end, there is a single-stranded DNA molecule called *bridge* (br) that complements concatenated linkers l_1 and l_2:

$$
\begin{array}{ll}
l_2|l_1 & 5' - \text{TCTACGCT}|\text{CGCAATTC} - 3' \\
br & 3' - \text{AGATGCGA GCGTTAAG} - 5'.
\end{array}
\qquad (6.15)
$$

Table 6.6 Encoding of the adjacency list of graph G'.

Region	Encoding
v_0	$5' - \text{ACTGACGGCCGTCAGT} - 3'$
v_1	$5' - \text{TACGATGGCCATCGTA} - 3'$
v_2	$5' - \text{GTGAGAGGCCTCTCAC} - 3'$
v_3	$5' - \text{CGTTCAGGCCTGAACG} - 3'$
v_4	$5' - \text{AGCTTCGGCCGAAGCT} - 3'$
v_1v_3	$5' - \text{TCACCTGGCCAGGTGA} - 3'$
v_1v_4	$5' - \text{GATCTGGGCCCAGATC} - 3'$
v_1v_5	$5' - \text{CTGTAAGGCCTTACAG} - 3'$
l_1	$5' - \text{CGCAATTC} - 3'$
l_2	$5' - \text{TCTACGCT} - 3'$
br	$5' - \text{GAATTGCGAGCGTAGA} - 3'$
sp	$5' - \text{TAAATAAATAAATAAA} - 3'$

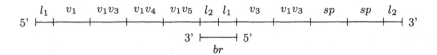

Fig. 6.44 Linked vertex templates for the vertices v_1 and v_3.

An excess of bridges is added to the solution of vertex templates. By annealing and ligation, partially double-stranded molecules are formed that correspond to subsets of vertices in G'; more precisely, multi-subsets as each vertex can occur more than once (Fig. 6.44). Afterwards, lambda exonuclease is added to the solution. This enzyme catalyzes the removal of strands with 5'-phosphate group. Thus, lambda exonuclease digests all bridges and all linked vertex templates whose first vertex template is phosphorylated. That is, the remaining non-digested molecules are single DNA strands that are linked vertex templates beginning with a non-phosphorylated vertex template.

In the second step, the linked vertex templates eventually form hairpins because of the palindromic encoding of the vertices and edges. That means if a linked vertex template contains a vertex or an edge at least twice, then a corresponding hairpin could be formed. These hairpin structures contain the recognition site of HaeIII and thus will be digested. Hence, the non-digested single DNA strands correspond to subsets of vertices in G' (no longer multi-subsets) that are non-adjacent by an edge in G'. These subsets correspond to independent sets in G' and therefore to cliques in G.

In the final step, the longest non-digested single DNA strands correspond to maximal independent sets in G' and therefore to maximum cliques in G. These strands are detected by gel electrophoresis.

The above routine was successfully implemented in the laboratory for the graph G shown in Figure 6.42. The experiments made use of dephosphorylated vertex templates for v_1 and v_2 yielding the maximum clique $\{v_2, v_3, v_4, v_5\}$.

6.3.5 Hairpin Structures

This section formalizes the notion of DNA hairpin secondary structure and examines its basic properties, based on work of L. Kari and coworkers (2006).

Let Σ be an alphabet. Let $\phi : \Sigma^* \to \Sigma^*$ be a morphic or anti-morphic involution, and let $k > 0$ be an integer. A string x in Σ^* is ϕ-k-*hairpin-free* or simply $\mathrm{hp}(\phi, k)$-*free* if $x = uyv\phi(y)w$ for some $u, v, w, y \in \Sigma^*$ implies $|y| < k$. It follows that all strings of length less than $2k$ are $\mathrm{hp}(\phi, k)$-free.

Let $\mathrm{hpf}(\phi, k)$ denote the set of all $\mathrm{hp}(\phi, k)$-free strings over Σ, and let $\mathrm{hp}(\phi, k)$ denote the complement of $\mathrm{hpf}(\phi, k)$ in Σ^*. Clearly, $\mathrm{hpf}(\phi, k)$ is a subset of $\mathrm{hpf}(\phi, k+1)$ for all $k > 0$. A language L over Σ is termed ϕ-k-*hairpin-free* or simply $\mathrm{hp}(\phi, k)$-*free* if L is a subset of $\mathrm{hpf}(\phi, k)$. It follows that a language L over Σ is $\mathrm{hp}(\phi, k)$-free if and only if $L \cap \Sigma^* y \Sigma^* \phi(y) \Sigma^* = \emptyset$ for all strings y of length at least k.

Example 6.20. Let Δ be the DNA alphabet, and let ϕ denote the reverse complementarity on Δ^*. The $\mathrm{hp}(\phi, 1)$-free strings over Δ cannot contain Watson-Crick pairs, that is,

$$\mathrm{hpf}(\phi, 1) = \{\mathrm{A}, \mathrm{C}\}^* \cup \{\mathrm{A}, \mathrm{G}\}^* \cup \{\mathrm{T}, \mathrm{C}\}^* \cup \{\mathrm{T}, \mathrm{G}\}^* .$$

On the other hand, the string $\mathrm{AATT} = \mathrm{AA}\phi(\mathrm{AA})$ belongs to $\mathrm{hpf}(\phi, 3)$, but not to $\mathrm{hpf}(\phi, 2)$. ◇

Theorem 6.21. *The languages* $\mathrm{hpf}(\phi, k)$ *and* $\mathrm{hp}(\phi, k)$ *are regular.*

Proof. Let ϕ be a morphic involution on Σ^*. Consider the non-deterministic finite state automaton in Figure 6.45, which accepts all strings of the form $uyv\phi(y)w$ with $|y| \geq 3$. This automaton can be easily modified so that it accepts all strings of the shape $uyv\phi(y)w$ with $|y| = 3$. The automaton is similarly defined if ϕ is an anti-morphic involution on Σ^*. A union of such automata yields a non-deterministic finite state automaton that accepts all strings of the form $uyv\phi(y)w$ so that $|y| < k$. Thus the language $\mathrm{hpf}(\phi, k)$ is regular. It follows that its complement $\mathrm{hp}(\phi, k)$ is also regular. □

Theorem 6.22 (Hairpin Freedom Problem). *The problem of deciding whether a regular language is* $\mathrm{hp}(\phi, k)$-*free is solvable in linear time.*

Proof. Let L be a regular language over Σ. By definition, L is $\mathrm{hp}(\phi, k)$-free if and only if $L \cap \mathrm{hp}(\phi, k) = \emptyset$. The latter problem is solvable in $O(|M_k| \times |M|)$ time (i.e., in linear time in $|M|$), where M is a finite state automaton accepting L and M_k is a finite state automaton accepting $\mathrm{hp}(\phi, k)$. □

Theorem 6.23 (Maximum Hairpin Freedom Problem). *The problem of deciding whether for an* $\mathrm{hp}(\phi, k)$-*free language L and a regular language L', there is a string $w \in L' \setminus L$ so that $L \cup \{w\}$ is* $\mathrm{hp}(\phi, k)$-*free is solvable in time proportional to the acceptance of L times the acceptance of L'.*

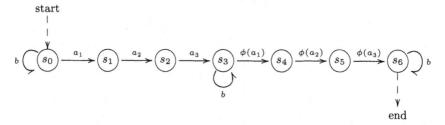

Fig. 6.45 Schematic representation of a non-deterministic finite state automaton for strings with hairpins of length at least 3.

Proof. Consider the word problem of deciding whether there is a string $w \in$ $\text{hpf}(\phi, k)$ so that $w \notin L$ and $w \in L'$. This problem is decidable in time $O(|M_k| \times |M| \times |M'|)$, where M is a finite state automaton accepting L, M' is a finite state automaton accepting L, and M_k is a finite state automaton accepting $\text{hpf}(\phi, k)$.   □

6.4 Computational Models

This section describes several autonomous DNA computational models that hold promise for performing complex information processing and control tasks in specific environments.

6.4.1 Neural Networks

First, the construction of a DNA model for neural networks based on the work of A. Mills, Jr. (2002) will be addressed.

Artificial Neural Networks

Artificial neurons with binary input and binary output were first devised by W. McCulloch and W. Pitts in 1943. A more general artificial neuron, called *perceptron*, was developed by F. Rosenblatt in 1958. A perceptron is a processing unit with n input signals and one output signal (Fig. 6.46). The perceptron linearly combines the input signals x_i via real-valued *weights* w_i, $1 \leq i \leq n$, to yield the *action potential*

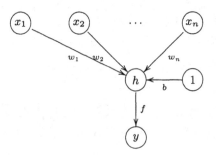

Fig. 6.46 Perceptron.

$$h = \sum_i w_i x_i + b , \qquad (6.16)$$

where b is a *bias*, a constant term that is independent of any input value. The action potential h is transformed by an *activation function* f to provide the output signal

$$y = f(h) . \qquad (6.17)$$

A perceptron basically implements a binary classifier. To see this, consider the simplest activation function providing the sign of the action potential h so that the input x is classified as either a positive or negative instance. The bias gives the perceptron a base level of activity. If the bias b is negative, then the action potential must have a positive value greater than b so that the input is classified as positive instance, and vice versa. Another common activation function is the S-shaped *sigmoid function* $f : h \mapsto 1/(1 + \exp(-h))$ (Fig. 6.47).

A *multilayer perceptron* (MLP) is a network of perceptrons (Fig. 6.48). An *L-layered MLP* consists of $L + 1$ ordered layers, the first layer is the *input layer*, the last layer the *output layer*, and the layers in between the *hidden layers*. The input layer simply consists of data storage units. Each non-input layer consists of a set of neurons, which receive their inputs from neurons of the previous layer and send their outputs to the subsequent layer. In this way, the input signal propagates through the network layer by layer from the input to the output layer. MLPs are typically fully connected (i.e., each neuron of one layer is connected via axons to all neurons in the subsequent layer). By choosing the right weights (or strengths of the axon connections), any continuous function can be approximated by an MLP with any given accuracy provided that a sufficient number of hidden neurons are available, as shown by K. Hornik and coworkers (1989).

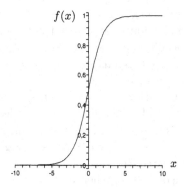

Fig. 6.47 Sigmoid function.

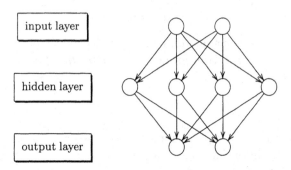

Fig. 6.48 Two-layered MLP.

MLPs are trained rather than programmed to carry out an information processing task. An MLP is trained using training samples whose classifications are known in advance. These samples are presented to the MLP. Each time the MLP gives the right (wrong) answer, it is either left invariant or reinforced (punished). There are several ways to reinforce or punish an MLP: changing weights (back propagation), adding or deleting connections, and adding or deleting hidden neurons. An MLP should be able to generalize from the training samples. To this end, the training samples should not be taught to the highest degree of accuracy. This requires that the number of training iterations must be carefully chosen. Otherwise, the MLP will be unable to extract important features.

DNA Perceptron Model

A neural network can be implemented by using a set of DNA oligonucleotides. In view of the perceptron, each input signal x_i and the output signal h are encoded by DNA oligonucleotides, and each weight or axon connection w_i is encoded by a partially double-stranded DNA molecule, $1 \leq i \leq n$. An axon-representing molecule is formed by using oligonucleotides that encode the desired input and output signals and temporarily attaching them to each other by a complementary linker oligonucleotide as in Adleman's first experiment. The two oligonucleotides are subsequently joined permanently by ligase. Finally, the output end of the linked oligonucleotide is protected from hybridization on its output end by extending the linker oligonucleotide along the output oligonucleotide via DNA polymerase (Fig. 6.49).

The network is operated by mixing input oligonucleotides and axon-representing oligonucleotides together and exposing this mixture to the action of DNA polymerase in a suitable reaction buffer. In this way, out-

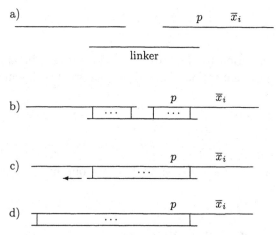

Fig. 6.49 A molecule representing an axon is formed in a preparation step. **a** Watson-Crick complementary input molecule with prefix p annealing to the linker, part of the Watson-Crick complementary output molecule, and linker molecule. **b** Annealing. **c** Ligation. **d** DNA polymerase extension.

put oligonucleotides are released by the DNA polymerase from those axon molecules that were primed on their output ends (Fig. 6.50).

A perceptron is implemented in a way that the input signals, axon connections, and output signals are represented by concentrations of the correspondingly encoded DNA molecules. In this way, the action potential (6.16) is to be interpreted as an equation of concentrations,

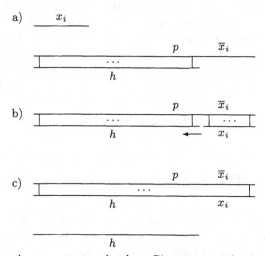

Fig. 6.50 Steps in forming an output molecule. **a** Given input molecule x_i and axon molecule formed in the preparation step. **b** Input molecule anneals to a corresponding axon molecule primed on the output end. **c** Resulting molecule is exposed to DNA polymerase leading to the release of the output oligonucleotide.

$$[h] = \sum_i [w_i][x_i] + [b] \ . \tag{6.18}$$

Moreover, the activation (6.17) is to be considered as a binary classifier so that a high (low) concentration of the action potential is classified as positive (negative) instance,

$$y = f([h]) \ . \tag{6.19}$$

This DNA perceptron model can be generalized to multi-layered networks by designing each layer individually and taking the output of one layer as input of the subsequent layer.

Gene Expression Profiling Diagnosis

Molecular diagnostic techniques for the classification of tumor cells are usually based on gene activity. The level of gene activity is represented by the concentrations of mRNA. However, RNA is not a very stable molecule and thus mRNA is converted by reverse transcriptase into complementary DNA (cDNA), which is more robust. The profile of active genes is then represented by their cDNA concentrations, which can be measured by using a DNA microarray. For this, the cDNA is stained with fluorescent dye and allowed to hybridize with an array of tens of thousands of DNA oligonucleotides representing many genes. The array is then exposed to light that excites the dye. The fluorescent intensities of the various cDNA strands are measured and compared with the intensities of the various cDNA oligonucleotides from a library of known cells. This approach allows the differentiation of pathological strains with indistinguishable phenotypes, which can be essential for determining the best therapy. The microarray framework is valuable for research, but for clinical use a simpler technique that gives answers on a shorter time-scale is required. MLPs are useful for classifying, generalizing and predicting based on a limited data set. Thus, MLPs might be helpful to tackle the expression profiling problem once the rules were established by careful laboratory and clinical studies.

6.4.2 Tic-Tac-Toe Networks

This section describes the construction of a DNA molecular automaton based on the work of M. Stojanovic and D. Stefanovic (2003), which encodes a version of the game of tic-tac-toe and interactively competes against a human opponent.

Molecular Logic Gates

The underlying molecular automaton is a Boolean network of logic gates that are made of allosterically modulated deoxyribozymes, and input and output are given by oligonucleotides. The logic gates are based on *molecular beacons* introduced by S. Tyagi and F. Kramer (1996), which are oligonucleotide probes able to report the presence of specific DNA in homogenous solutions. Molecular beacons are hairpin-shaped molecules with a quencher (R) at the 3' end and a fluorosphore (F) at the 5' end. These molecules are non-fluorescent as the loop keeps the quencher close to the fluorophore. However, when the oligonucleotide probe sequence in the loop hybridizes to a target DNA molecule forming a rigid double helix, the quencher is separated from the fluorophore restoring fluorescence (Fig. 6.51).

Deoxyribozymes are nucleic acid catalysts made of deoxyribonucleic acid. Examples of deoxyribozymes are phosphodiesterases, which cleave other oligonucleotides with shorter products as output, and ligases, which combine two oligonucleotides into a larger product. The molecular gates are based on phosphodiesterase E6, which cleaves an input fluorogenic oligonucleotide upon binding and releases the cleft oligonucleotides as output. This produces an increase in fluorescence by fluorescein (F) as the quencher (R) is separated (Fig. 6.52).

The YES gate is comprised of a deoxyribozyme E6 module and a hairpin module complementary to input oligonucleotide x. If the input oligonucleotide hybridizes to the hairpin module, the gate enters the active state, increasing fluorescence (Fig. 6.53).

The NOT gate is constructed by extending the non-conserved loop of the E6 core with a hairpin structure that is complementary to input oligonucleotide x. If the input oligonucleotide hybridizes to the hairpin module, the gate enters the inactive state not increasing fluorescence (Fig. 6.54).

The AND gate is obtained from the E6 core module by extending both ends with hairpin modules, one complementary to input oligonucleotide x and the other complementary to input oligonucleotide y (Fig. 6.55).

The AND-NOT gate realizes the Boolean function $(x, y) \mapsto x\bar{y}$ and is derived from the E6 core module by extending one end with a hairpin module,

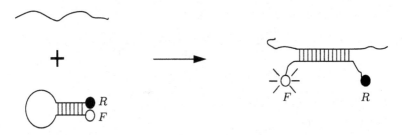

Fig. 6.51 Operation of molecular beacon.

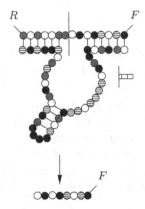

Fig. 6.52 Activity of phosphodiesterase E6. Imaginary: open circle = A, black circle = T, gray circle = C, and shaded circle = G.

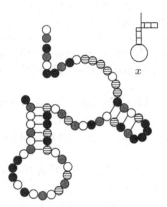

Fig. 6.53 YES gate.

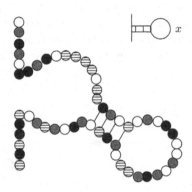

Fig. 6.54 NOT gate.

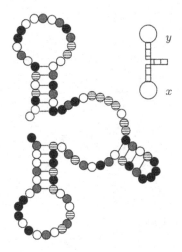

Fig. 6.55 AND gate.

which is complementary to input oligonucleotide x, and extending the loop of the E6 core with a hairpin structure complementary to input oligonucleotide y (Fig. 6.56).

The AND-AND-NOT gate implements the Boolean function $(x, y, z) \mapsto xy\bar{z}$ and is derived from the E6 core module by extending both ends with hairpin modules, one complementary to input oligonucleotide x and the other complementary to input oligonucleotide y, and extending the loop of the E6 core with a hairpin structure complementary to input oligonucleotide z (Fig. 6.57).

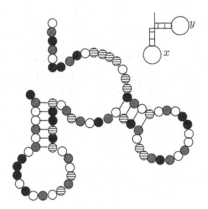

Fig. 6.56 AND-NOT gate.

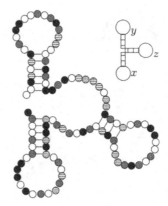

Fig. 6.57 AND-AND-NOT gate.

Game Play

Tic-tac-toe is a two-player game played on a grid of 3×3 squares (Fig. 6.58). The two players take turns claiming squares and marking them, X for one player and O for the other. No player can claim a square that has already been claimed. A player wins the game if she claims three squares in a row, column, or diagonally. The game ends with a draw if there is no winner after the players have claimed all nine squares. Wins or draws are favorable outcomes of the game.

A game-winning strategy is illustrated in Figure 6.59. This strategy suggests that the machine goes first placing the mark into the center (square 5), and by symmetry, the (human) opponent places the mark either into square 1 or 4. No further restrictions are made on the subsequent turns. There are nineteen games: ten end in victory for the machine after two moves of the human, seven after three moves, one after four moves, and one game is a draw.

The game strategy is implemented by Boolean functions that correspond one-to-one to the squares. The variables are associated with the opponent's plays so that a variable is assigned the value "true" if the human places a mark into the corresponding square. A function evaluates to "true" when the

1	2	3
4	5	6
7	8	9

Fig. 6.58 Game board.

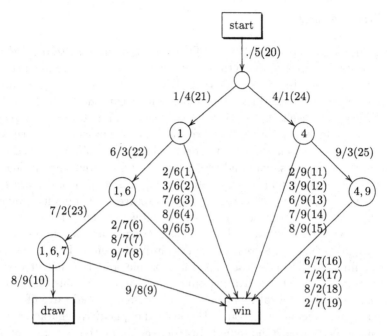

Fig. 6.59 Game strategy: The machine goes first. Each node is labelled by the squares marked by the human, and each edge is labelled $i/o(n)$, where i is the mark of the human, o is the response of the machine, and n is an integral edge label.

machine can place a mark into the square corresponding to this function. This requires that $f_5 = 1$ so that at the beginning of the game, the machine can place a mark into square 5.

For instance, the machine places a mark into square 2 on the edges 17, 18, and 23 (Fig. 6.59). Thus, $f_2 = x_4x_9x_7 + x_4x_9x_8 + x_1x_6x_7$. However, these functions are ambiguous (i.e., two or more functions can evaluate to "true" during the game while the strategy always provides a single outcome). These ambiguities can be resolved by introducing appropriate negated variables into the functions. This leads to a set of logically equivalent functions in which all monomials can be realized by molecular gates:

$$f_1 = x_4$$
$$f_2 = x_6x_7\overline{x}_2 + x_7x_9\overline{x}_1 + x_8x_9\overline{x}_1$$
$$f_3 = x_1x_6 + x_4x_9$$
$$f_4 = x_1$$
$$f_5 = 1$$
$$f_6 = x_1x_2\overline{x}_6 + x_1x_3\overline{x}_6 + x_1x_7\overline{x}_6 + x_1x_8\overline{x}_6 + x_1x_9\overline{x}_6$$
$$f_7 = x_2x_6\overline{x}_7 + x_6x_8\overline{x}_7 + x_6x_9\overline{x}_7 + x_2x_9\overline{x}_1$$
$$f_8 = x_7x_9\overline{x}_4$$
$$f_9 = x_7x_8\overline{x}_4 + x_2x_4\overline{x}_9 + x_3x_4\overline{x}_9 + x_4x_6\overline{x}_9 + x_4x_7\overline{x}_9 + x_4x_8\overline{x}_9 .$$

In Vitro Gaming

The game is played in the 3×3 wells of a 386-well plate termed MAYA, which hosts 24 types of deoxyribozymes in nine wells. There are two YES gates, two AND gates, and 19 ANDANDNOT gates; and one permanently active deoxyribozyme in the center square, with a concentration of 200 nM of the YES gates in wells 1 and 4, and a concentration of 150 nM of the other gates. The game starts by adding a divalent metal ion cofactors, like Mg^{2+} ions, to all wells, which activates the deoxyribozymes. MAYA places the first mark into the center square and thus there are eight input oligonucleotides, one for each of the remaining squares. The input oligonucleotides are 15- or 17-mers that exhibit no strong secondary structures. The opponent makes a move by adding an input oligonucleotide to all wells setting the corresponding logical variable to "true". This addition triggers an output response by the machine given by the greatest increase in fluorescence. The output interface is a fluorescence plate reader, and fluorogenic activity is detected within 15 min from the time the activating inputs were added. For this, the gates are optimized to achieve a ratio of at least 10:1 between initial rates of cleavage in the active versus inactive states. However, the quantitive characteristics of gates may vary for each individual gate, depending on the structure of input oligonucleotides and the concentration of individual components of the reaction. The authors reported that in a total of more than one hundred games played, no erroneous moves have been detected.

The deoxyribozyme-based automaton MAYA consists of individual biomolecular building blocks that behave as feed-forward circuits carrying out Boolean computations without human interface-operated steps. The output of one gate may be used as the input of another gate. Hence, automata are conceivable that can carry out more complex tasks. Those automata might be useful in synthetic biology to learn about genetic, regulatory, and metabolic networks.

6.4.3 Logic Circuits

This section addresses the construction of enzyme-free DNA logic circuits based on the work of E. Winfree and coworkers (2006), which embodies basic digital design principles such as Boolean logic, cascading, restoration, and modularity. The constructions rely on DNA reactions that can be driven without enzyme or (deoxy)ribozyme catalysis, first studied by A.J. Turberfield and coworkers (2003).

Toehold Kinetics

The key phenomenon for the implementation of DNA logic gates is strand displacement via branch migration mediated by toeholds (Fig. 6.60). For this, consider the reaction

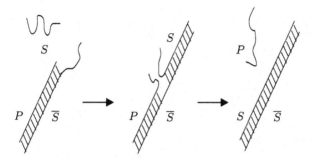

Fig. 6.60 Strand displacement by branch migration.

$$P\overline{S} + S \rightarrow S\overline{S} + P \ . \tag{6.20}$$

This reaction is thermodynamically favorable, because of the additional base pairs that are formed. The kinetics of this reaction depends on the length of the single-stranded overhang in $P\overline{S}$, known as a *toehold*. The single strand S initially binds to the partial double strand $P\overline{S}$ at the toehold. The longer the toehold, the more likely the reaction will enter a *branch migration* phase prior to dissociation. Branch migration comprises isoenergetic steps in which the final base pair of P to $\overline{S}$ is replaced by a base pair of S to $\overline{S}$. When the branch point reaches the end of the complex, the strand P dissociates. This step is irreversible as there is no toehold for P. This biochemical reaction is accomplished by DNA without the assistance of any enzymes.

DNA Logic Gates

DNA logic gates are entirely determined by base pairing and breaking, and the logical values 0 and 1 are represented by low and high concentrations, respectively. Each logic gate consists of one or more gate strands and one output strand. The output strand serves either as an input to a downstream gate or is modified with a dye label to provide the readout in a fluorescence experiment.

An AND gate is given by a partially double-stranded DNA molecule consisting of three DNA strands, E_{out} (57 nt), F (60 nt), and G (36 nt). The gate contains three toehold binding regions (6 nt), one at the 3' end of G, one between E_{out} and G, and one inside of F. The input strands F_{in} and G_{in} (36 nt) are complementary to the respective strands F and G within the gate. When input strands are given into the solution containing the gate, the computation is initiated (Fig. 6.61). The input strand G_{in} binds to the toehold at the 3' end of G and displaces the first gate strand by toehold kinetics. This exposes the toehold for the subsequent input strand and releases an

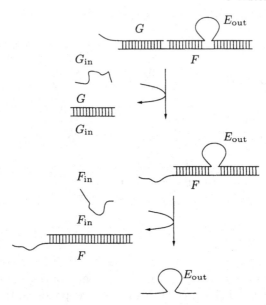

Fig. 6.61 Implementation of AND gate by toehold kinetics.

inert double-stranded waste product. A similar process takes place for the second input F_{in}. Hence, an output strand is released if and only if both input strands are present. The output strand may serve as an input strand for another logic gate.

A single-input AND gate can be used as a *translator gate*, which converts a signal encoded by the input strand to a signal encoded by the output strand.

An OR gate can be implemented by three translator gates so that two translator gates take the input signal and release the same output signal, while the third translator gate uses such an output signal to provide the output signal of the gate. Hence, an output strand is released if and only if at least one input strand is present.

A NOT gate consists of a translator gate and an inverter strand. If the input strand is present, the inverter strand preferably hybridizes to it. Otherwise, the inverter strand triggers the translator gate and thus provides an output signal. Inverter strands must be simultaneously added to the input of the circuit and therefore NOT gates are restricted to the first layer of the circuit. By De Morgan's law, this is sufficient to realize arbitrary Boolean functions.

DNA Logic Circuits

The described DNA gates provide a full set of logic gates using short oligonucleotides as input and output. As input and output are of the same type, the gates can be organized in cascades to implement logic circuits.

Signal restoration eventually becomes necessary when a gate fails to produce enough output strands when triggered, or when a gate leaks by spontaneous release of an output strand. The first type of error requires increasing a moderate output amount to a full activation level, while the second type of error requires decreasing a small output amount to a negligible level. For this, a signal restoration module providing thresholding and amplification should be incorporated into large circuits at multiple intermediate points to ensure that the digital signal representation remains stable.

Signal thresholding can be realized by *threshold gates*, which are three-input AND gates with identical first and third input strands. A small amount of input will cause most gates to lose only their first and second gate strands and thus release no output, while input concentrations two times higher than the concentration of threshold gates will cause most gates to produce output. The threshold gate's concentration sets the threshold for a sigmoidal nonlinearity. But the threshold gate's output cannot exceed half the input signal and thus subsequent amplification is necessary.

Signal amplification can be based on feedback logic. An *amplifier gate* is a two-gate feedback circuit amplifying the fluorescence output signal without producing an output strand. The circuit consists of two translator gates so that the output of the first acts as input for the second and the output of the second acts as input for the first. This fluorescence amplifier linearly amplifies the output signal with time.

The overall approach is modular and scalable and allows interfacing with predesigned subcircuits. The design was experimentally tested by a larger circuit of eleven gates, with six inputs given by DNA analogs of mouse microRNAs. The circuits may work well with RNA inputs instead of DNA inputs, because the gate functions solely depend on Watson-Crick complementarity. Potential applications of DNA logic circuits are to control nanoscale devices in vitro and to detect complex expression patterns in situ.

6.4.4 Turing Machines

This section describes an autonomous DNA Turing machine based on the work of J.H. Reif and coworkers (2005). This DNA nanomechanical device simulates the general operation of an arbitrary 2-state 5-symbol Turing machine, whose head moves either to the left or to the right in every transition. Thus, the autonomous DNA Turing machine can implement the universal Turing machine given in Example 2.33.

Data Representation

The autonomous Turing machine consists of two major components: two parallel arrays of dangling molecules tethered to two rigid tracks (Fig. 6.62).

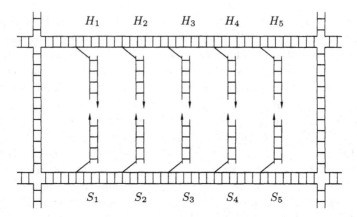

Fig. 6.62 Schematic illustration of an autonomous DNA Turing machine.

The two rigid tracks can be implemented as a rigid DNA lattice, which can be made from a diverse set of branched DNA molecules such as DX molecules.

A *dangling molecule* is a double-stranded DNA molecule with sticky ends, one of which is linked to the track by a flexible single strand extension. The upper and lower arrays of dangling molecules are called *head molecules*, denoted by H, and *symbol molecules*, denoted by S, respectively. The dangling molecules can move rather freely so that a dangling molecule can interact with its neighboring dangling molecules along the same track or the dangling molecule immediately below or above it. This can be achieved by the rigidity of the tracks and the proper spacing of the dangling molecules. The array of head molecules represents the moving head of the Turing machine, while the array of symbol molecules provides the data tape of the Turing machine.

The design involves another type of molecules, *floating molecules*, which freely float in the solution. Floating molecules are double-stranded DNA molecules with one sticky end, which either specify the computational transitions (rule molecules) or assist in carrying out the machine operations (assisting molecules).

A dangling molecule is encoded by the string $X^a[y]^b$ or $[y]^b X^a$, where X is the double-stranded portion, $[y]$ is its sticky end portion, a is the state, symbol, or position information encoded in X, and b is the state, symbol, or position information encoded in $[y]$. At the beginning of the computation, the array of head molecules along the head track is configured as

$$\ldots ([\bar{h}]^{p_{i-1}} H_{i-1}^{p_{i-1}})(\hat{H}_i^{p_i q}[s])([\bar{h}]^{p_{i+1}} H_{i+1}^{p_{i+1}}) \ldots \qquad (6.21)$$

The ith head molecule (marked by a hat) represents the current position of the machine head. It encodes the current state q in its double-stranded portion and possesses a sticky end $[s]$ that is complementary to the sticky end $[\bar{s}]$ of the symbol molecule below it. Furthermore, the position type information

p_i is encoded both in the sticky end portion and the double-stranded portion of the head molecules. This will guarantee that the head array has the shape (6.21) after each computational step. The array of symbol molecules along the symbol track provides the Turing tape and is given as

$$([\bar{s}]S_1^{c_1})([\bar{s}]S_2^{c_2})([\bar{s}]S_3^{c_3})\ldots \tag{6.22}$$

All symbol molecules possess sticky ends $[\bar{s}]$ and the symbols c_i are encoded by the double-stranded portions.

Computation

The one-step behavior of the Turing machine given by the transition $\delta(q, c) = (q', c', p')$ is implemented in eight steps. In the first step, the current head $\hat{H}_i^{pq}$ (encoding position type p and current state q) and the symbol molecule $[\bar{s}]S_i^c$ (encoding current symbol c) below it hybridize and are ligated into $(H_iS_i)^{pqc}$. This ligation product is cut by an endonuclease into $\hat{H}_i^p[r]^{qc}$ and $[\bar{r}]^{qc}S_i$ so that the sticky ends of both $\hat{H}_i$ and S_i encode the current symbol and state:

$$\hat{H}_i^{pq}[s] + [\bar{s}]S_i^c \to (H_iS_i)^{pqc} \to \hat{H}_i^p[r]^{qc} + [\bar{r}]^{qc}S_i . \tag{6.23}$$

In the second step, a rule molecule $R[r]^{qc}$ hybridizes and is ligated with symbol molecule $[\bar{r}]^{qc}S_i$. This product is cut into the waste molecule $R_w^{qc}[e]^{c'}$ and the symbol molecule $[\bar{e}]^{c'}S_i$ so that the sticky end $[\bar{e}]$ encodes the new symbol c':

$$R[r]^{qc} + [\bar{r}]^{qc}S_i \to (RS_i)^{qcc'} \to R_w^{qc}[e]^{c'} + [\bar{e}]^{c'}S_i . \tag{6.24}$$

In the third step, the symbol molecule $[\bar{e}]^{c'}S_i$ is restored to its default configuration $[\bar{s}]^{c'}S_i$ so that it can eventually interact with the head molecule H_i. For this, an assisting molecule $E^{c'}[e]^{c'}$ hybridizes and is ligated with symbol molecule $[\bar{e}]^{c'}S_i$. The resulting product is cut, yielding the waste molecule $E_w[s]$ and the symbol molecule $[\bar{s}]S_i^{c'}$ encoding the symbol c' in its double-stranded portion:

$$E^{c'}[e]^{c'} + [\bar{e}]^{c'}S_i \to (ES)^{c'} \to E_w[s] + [\bar{s}]S_i^{c'} . \tag{6.25}$$

The waste molecule $E_w[s]$ may hybridize and be ligated with some symbol molecule $[\bar{s}]S_l$. This may decrease the efficiency of the computation when the concentration of the waste molecules increases. However, this is counteracted by the endonuclease used in the first step, which may subsequently cut such ligation products.

In the fourth step, the head molecule $\hat{H}_i^p[r]^{qc}$ hybridizes and is ligated with a rule molecule $[\bar{r}]^{qc}R$. The ligation product is cut into waste molecule

$[\bar{h}]^{p'} R_w$ and head molecule $\hat{H}_i^{pq'}[h]^{p'}$, which encodes the new state q' in the double-stranded portion and the motion direction p' in the sticky end portion:

$$\hat{H}_i^p[r]^{qc} + [\bar{r}]^{qc}R \to (H_iR)^{pqc} \to \hat{H}_i^{pq'}[h]^{p'} + [\bar{h}]^{p'} R_w . \qquad (6.26)$$

In the fifth step, the head molecule $\hat{H}_i^{pq'}[h]^{p'}$ hybridizes and is ligated to either its left or right neighbor $[\bar{h}]^{p'} H_j'$, $j = i - 1$ or $j = i + 1$, depending on the sticky end information p'. The ligation product is cut into head molecules $H_i^p[t]^{pp'q'}$ and $[t]^{pp'q'} \hat{H}_j'$ with $j = i - 1$ or $j = i + 1$ so that both sticky ends encode position type p of H_i, position type p' of H_j, and new state q':

$$\hat{H}_i^{pq'}[h]^{p'} + [\bar{h}]^{p'} H_j' \to (H_iH_j)^{pp'q'} \to H_i[t]^{pp'q'} + [t]^{pp'q'} \hat{H}_j' . \qquad (6.27)$$

In the sixth step, the head molecule $\hat{H}_j'$ is altered similar to the fourth step so that it can interact with a symbol molecule:

$$\hat{H}_j'[t]^{pp'q'} + [t]^{pp'q'}T \to (H_j'T)^{pp'q'} \to \hat{H}_j'^{q'}[s] + [\bar{s}]T_w . \qquad (6.28)$$

In the seventh step, the sticky end of the head molecule H_i is modified by an assisting molecule in order to provide a new sticky end:

$$[t]^{pp'q'} H_i + T[t]^{pp'q'} \to (TH_i)^{pp'q'} \to T_w[g]^{pp'q'} + [\bar{g}]^{pp'q'} H_i . \qquad (6.29)$$

In the last step, the head molecule H_i is grown by assisting molecules in a series of alternating hybridizations, ligations, and cleavages in order to restore its default state:

$$[\bar{g}]^{pp'q'} H_i \to [\bar{h}]^p H_i^p . \qquad (6.30)$$

The autonomous DNA Turing machine implements universal computation and thus provides universal translational motion given by the motion of the head symbol relative to the tracks.

The design of the autonomous DNA Turing machine was verified by a computer simulation making use of four endonucleases. A full experimental implementation of this machine requires the tackling of two major technical issues. The first issue is to accommodate the *futile reactions* that take place during the operational cycle. Futile reactions eventually occur between a rule molecule $R[r]^{qc}$ and its complement $[\bar{r}]^{qc}R$, a head molecule $\hat{H}^{pq}[s]$ and a symbol molecule $[\bar{s}]S^c$, or a rule molecule $R[r]^{qc}$ and a symbol molecule $[\bar{r}]^{qc}S$, among others. These reactions decrease the efficiency of the design, but can be made fully reversible so that they do not block, reverse, or alter the operability. The second issue is the limited encoding of symbols, states, and positions as DNA words, which is dictated by the sizes of the recognition, restriction, and spacing regions of endonucleases. For this, the sticky ends must be carefully selected to avoid undesirable reactions. This may be achieved by extending the DNA alphabet with unnatural bases.

Concluding Remarks

Autonomous DNA computational models can make use of structural DNA nanotechnology, which aims at a rational approach to the construction of new biomaterials, such as individual geometrical objects and nanomechanical devices, including extended constructions such as periodic matter. In the last decade, DNA was shown to be capable of fulfilling all of these roles in prototype systems. A key element in this work are DX molecules, which not only serve as model systems for structures proposed to be involved in genetic and meiotic recombination, but also allow the implementation of small cellular automata or other computational models in a molecular context.

Biological organisms perform complex information processing and control tasks using sophisticated biochemical circuits. To date, no man-made biochemical circuits even remotely approach the complexity and reliability of silicon-based electronics. Once principles for their design are established, circuits could be used to control nanoscale devices in vitro, to interface with existing biological circuits in vivo, or to analyze complex chemical samples in situ. In particular, rational design of DNA devices is simplified by the predictability of Watson-Crick base pairing. Consequently, DNA devices could be a promising alternative to proteins for synthetic chemical circuits. The remaining challenge is to design chemical logic gates that can be combined to construct large reliable circuits.

References

1. Adar R, Benenson Y, Linshiz G, Rosner A, Tishby N, Shapiro E (2004) Stochastic computing with biomolecular automata. Proc Natl Acad Sci USA 101:9960–9965
2. Benenson Y, Paz-Elizur T, Adar R, Keinan E, Livneh Z, Shapiro E (2001) Programmable and autonomous computing machine made of biomolecules. Nature 414:430–434
3. Benenson Y, Paz-Elizur T, Adar R, Livneh Z, Shapiro E (2003) DNA molecules providing a computing machine with both data and fuel. Proc Natl Acad Sci USA 100:2191–2196
4. Condon A (2004) Automata make antisense. Nature 429:361–362
5. Durbin R, Eddy SR, Krogh A, Mitchinson G (1998) Biological sequence analysis: probabilistic models of proteins and amino acids. Cambridge Univ Press, Cambridge
6. Fu TJ, Seeman NC (1993) DNA double-crossover molecules. Biochem 32: 3211–3220

7. Hagiya M, Arita M (1998) Towards parallel evaluation and learning of Boolean μ-formulas with molecules. Proc 3rd DIMACS Workshop DNA Based Computers, 57–72

8. Hagiya M, Sakamoto M, Arita M, Kiga D, Yokoyama S (1997) Towards parallel evaluation and learning of boolean u-formulas with molecules. Proc 3rd DIMACS Workshop DNA Based Computers, 105–114

9. Hornik K, Stinchcombe, White H (1989) Multilayer feedforward networks are universal approximators. Neural Networks 2:359–366

10. Kari L, Konstantinidis S, Losseva E, Sosik P, Thierrin G (2006) Hairpin structures in DNA words. LNCS 3492:158–170

11. Kroeker WD (1976) Mung bean nuclease I: terminally directed hydrolisis of native DNA. Biochem 15:4463–4467.

12. Kuramochi J, Sakakibara Y (2006) Intensive in vitro experiments of implementing and executing finite automata in test tube. LNCS 3892:193–202

13. Li X, Yang X, Qi J, Seeman NC (1996) Antiparallel DNA double crossover molecules as components for nanoconstruction. J Am Chem Soc 1118:6131–6140

14. Martínez-Pérez I, Zhong G, Ignatova Z, Zimmermann KH (2005) Solving the Hamiltonian path problem via DNA hairpin formation. Int J Bioinform Res Appl 1:389–398

15. Martinez-Perez I, Zhong G, Ignatova Z, Zimmermann KH (2007) Computational genes: a tool for molecular diagnosis and therapy of aberrant mutational phenotype. BMC Bioinform 8:365

16. McCulloch WS, Pitts W (1943) A logical calculus of the ideas immanent in nervous activity. Bull Math Biophys 7:115-133

17. Mills Jr AP (2002) Gene expression profiling diagnosis through DNA molecular computation. Trends Biotechnol 20:137–140

18. Moore EF (1962) Sequential machines: selected papers. Addison-Wesley Reading MA

19. Reif JH, Sahu S, Yin P (2005) Compact error-resilient computational DNA tiling assemblies. LNCS 3384:293–307

20. Reif JH, Sahu S, Yin P (2006) Complexity and graph self-assembly in accretive systems and self-destructible systems. LNCS 3892:257–274

21. Rothemund PWK (1996) A DNA and restriction enzyme implemenation of Turing machines. Proc 1st DIMACS Workshop DNA Based Computers 75–119

22. Sahu S, Reif JH (2007) Capabilities and limits of compact error-resilience methods for algorithmic self-assembly in two and three dimensions. LNCS 4287:223–238

23. Rosenblatt F (1958) The perceptron: a probabilitistic mode for infomation storage and processing in the brain. Psychological Rev 65:386–408

24. Sakakibara Y, Suyama A (2000) Intelligent DNA chips: logical operation of gene expression profiles on DNA computers. Gen Inform 11:33-42

25. Sakamoto K, Gouzu H, Komiya K, Kiga D, Yokoyama S, Yokomori T, Hagiya M (2000) Molecular computation by DNA hairpin formation. Science 288: 1223–1226.

26. Sakamoto K, Gouzu H, Komiya K, Kiga D, Yokoyama S, Ikeda S, Sugiyama H, Hagiya M (1999) State transitions by molecules. Biosystems 52:81–91

27. Schwacha A, Kleckner N (1995) Identification of double Holliday junctions as intermediates in meiotic recombination. Cell 83:783–791

28. Seelig G, Soloveichnik D, Zhang DY, Winfree E (2006) Enzyme-free nucleic acid logic circuits. Science 314:1585–1588

29. Seelig G, Yurke B, Winfree E (2006) Catalyzed relaxation of a metastable DNA fuel. J Am Chem Soc 128:12211–12220

30. Seeman NC (1982) Nucleic acid junctions and lattices. J Theor Biol 99:237–247

31. Seeman NC (1990) De novo design of sequences for nucleic acid structural engineering. J Biomol Struct Dyn 8:573–581

32. Seeman NC (1993) DNA double-crossover molecules. Biochem 32:3211–3220
33. Seeman NC (1999) DNA engineering and its application to nanotechnology. Trends Biotechnol 17:437–443
34. Stojanovic MN, Stefanovic D (2003) A deoxyribozyme-based molecular automaton. Nat Biotechnol 21:1069–1074
35. Thaler DS, Stahl FW (1988) DNA double-chain breaks in recombination of phage and of yeast. Annu Rev Genet 22:169–197
36. Tyagi S, Kramer FR (1996) Probes that fluoresce upon hybridization. Nature Biotech 14:303–308
37. Winfree E (1998) Algorithmic self-assembly of DNA. PhD thesis, Caltec, Pasadena
38. Winfree E, Liu F, Wenzler LA, Seeman NC (1998) Design and self-assembly of two-dimensional DNA crystals. Nature 394:539–544
39. Winfree E, Yang X, Seeman NC (1996) Universal Computation via self-assembly of DNA: some theory and experiments. Proc 2nd DIMACS Workshop DNA Based Computers 191-213
40. Turberfield AJ, Mitchell JC, Yurke B, Mills Yr AP Blakey MI, Simmel FC (2003) DNA fuel for free-running nanomachines. Phys Rev Lett 90:118102
41. Yin P, Turberfield AJ, Sahu S, Reif JH (2005) Design of autonomous DNA nanomechanical device capable of universal computation and universal translational motion. LNCS 3892:399–416
42. Yurke B, Mills Jr AP (2003) Using DNA to power nanostructures. Genet Program Evolable Machines 4:111–122

Chapter 7
Cellular DNA Computing

Abstract Cellular DNA computing investigates computational properties of DNA in its natural environment: the living cell. This chapter reviews some recent DNA computing models which are proposed to work at the cellular level. The first model describes gene synthesis during sexual reproduction in ciliates, while the other models focus on logical control or manipulation of cellular expression patterns.

7.1 Ciliate Computing

Ciliates are a diverse group of unicellular eukaryotic organisms. The process of gene assembly in ciliates is one of the most complex instances of DNA computation known in living organisms. This section provides an introduction to this process from the computational point of view.

7.1.1 Ciliates

Ciliates form a diverse group of unicellular organisms that are found in practically all water-rich environments. The pond water critters are an ancient group of eukaryotes with about 7,000 known species. All ciliates have hundreds of tiny hair-like cilia which beat in unison to propel the organism through water and to sweep food down into their oral apparatus (Fig. 7.1). Some ciliates are very small, not much larger than the largest bacteria. Others, like the "trumpet animalcule" *Stentor*, can become two millimeters long.

Ciliates possess organelles that help to perform all types of physiological activities. They have developed the unique feature of nuclear dualism. They possess two nuclei that are functionally different: the micronucleus and the

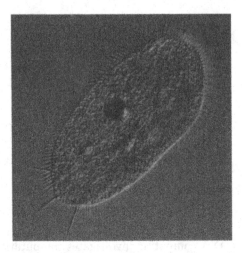

Fig. 7.1 Ciliate belonging to the genus *Stylonychia*. Image courtesy of BioMEDIA Associates (www.ebiomedia.com).

macronucleus. The *micronucleus* contains a diploid, meiotic germline genome whose genes are not directly expressed. For instance, the micronuclear genome of the ciliate protist *Sterkiella histriomuscorum* (formerly *Oxytricha trifallax*) has the size of approximately 5×10^8 bp. The genome consists of about 20,000 to 30,000 genes that are organized in about 100 micronuclear chromosomes.

The *macronucleus*, a highly specialized expression organelle, is a somatic nucleus providing RNA transcripts needed for the vegetative functioning of the cell. In *Sterkiella* the macronuclear genome consists of genes deployed on a collection of about 20,000 different miniature (gene-sized) chromosomes, each at a ploidy of about 1,000 per macronucleus. In this way, a single macronucleus contains more than 20 million DNA molecules. These molecules range from 250 bp to 40 kbp and are protected from endonuclease digestion by a telomerase binding protein that forms a tight complex with a telomere sequence appended at both ends of the molecule. This telomere sequence is 36 bp long and added by telomerase to both ends in the repeated pattern: GGGG alternating with TTTT.

The somatically active macronucleus adheres its genetic material from the germline micronucleus after sexual reproduction (conjugation) (Fig. 7.2). Under certain conditions, such as overcrowding or environmental stress, ciliates proceed to sexual reproduction, and during this process micronuclear genes are converted into their macronuclear form. This conversion is called *gene assembly*. The process of gene assembly is very intricate, since the micronuclear and the macronuclear forms of the same gene may be drastically different. Indeed, gene assembly in ciliates belongs to the most sophisticated DNA processing known so far in living organisms. During sexual reproduction two ciliate cells stick together and form a connected cytoplasmic channel.

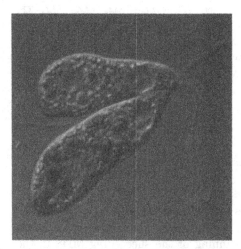

Fig. 7.2 Ciliates in conjugation. Image courtesy of BioMEDIA Associates (www.ebiomedia.com).

At the same time, the micronuclei undergo meiosis. In meiosis, the diploid micronuclei divide twice to produce four haploid descendent micronuclei, each of which contains one copy of each chromosome. Both cells exchange one haploid micronucleus through the cytoplasmic channel. The migrating haploid micronucleus fuses with a stationary haploid micronucleus in each cell and in this way forms a new diploid micronucleus. Then the two cells separate, and the new diploid micronucleus in each cell divides into two (without cell division). One of the descendent nuclei develops into a new macronucleus in each cell during the next few days. At the same time, the old macronucleus and the unused haploid micronuclei are destroyed.

As is common in eukaryotes, the micronuclear genome mostly consists of spacer DNA separating the genes, and the genes are interrupted by non-coding segments termed *internally eliminated sequences* (IESs). The gene segments interrupted by IESs are protein-coding and are called *macronuclear destined sequences* (MDSs). Almost all *Spichotrich* and *Paramecium* genes carry multiple IESs. In the case of micronuclear genes interruped by IESs, contiguous coding segments may neither be adjacent nor on the same strand. This means that the order of the MDSs in a micronuclear gene may be a permutation of the natural ordering in the macronuclear gene, and some MDSs may lie on the complementary strand (i.e., they are inverted in the micronuclear gene) (Fig. 7.3). In the micronuclear genome of *Sterkiella*, as many as 25–30% of the genes appear to be scrambled in complex patterns ranging from a few to over 50 coding segments, present in one or more micronuclear loci. Gene assembly removes IESs by precise excision and properly reorders and joins these fragments into the correct linear gene-sized, translatable order and orientation (Fig. 7.4). To this end, characteristic short sequences at the

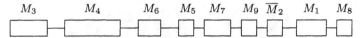

M_3 M_4 M_6 M_5 M_7 M_9 $\overline{M}_2$ M_1 M_8

Fig. 7.3 The actin I gene in the micronucleus of *S. nova*. The MDS sequences are given as rectangles (with MDS M_2 inverted) and the interspersed IESs are shown as line segments.

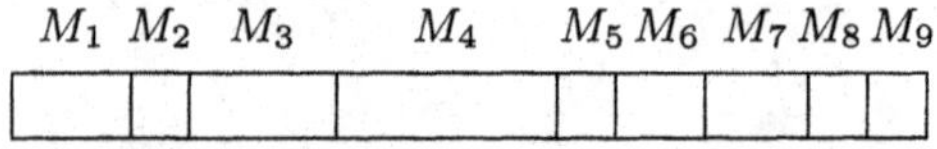

M_1 M_2 M_3 M_4 M_5 M_6 M_7 M_8 M_9

Fig. 7.4 The actin I gene in the macronucleus of *S. nova*.

ends of the MDSs termed *pointers* play a central role in gene assembly. The pointer at the end of an MDS coincides as a nucleotide sequence with the pointer at the beginning of the succeeding MDS in the macronuclear gene (Fig. 7.5). In this way, the pointers guide the recombination of the MDSs into a macronuclear gene (Fig. 7.6). Finally, the released DNA molecules replicate many times so that at the end of the macronuclear development each gene is present in about 1,000 copies of individual molecules.

7.1.2 Models of Gene Assembly

Two models for gene assembly in ciliates have been recently proposed. The first model, devised by L. Kari and L. Landweber (1999), is intermolecular and consists of the following three operations:

$5' - \ldots$TCGATCGG|ACATTC|aacattgaatctaat|ACATTC|GATCTAGGT$\ldots - 3'$
$3' - \ldots$AGCTAGCC|TGTAAG|aacattgaatctaat|TGTAAG|CTAGATCCA$\ldots - 5'$

Fig. 7.5 A section of the micronuclear gene encoding βTP in *S. histriomuscorum*. This section contains in turn a postfix of MDS 2, IES 2, and a prefix of MDS 3. The two repeated MDS subsequences (pointers) are ACATTC/TGTAAG, and the IES is in lower case letters. Vertical bars indicate subsequences.

$5' - \ldots$TCGATCGG|ACATTC|GATCTAGGT$\ldots - 3'$
$3' - \ldots$AGCTAGCC|TGTAAG|CTAGATCCA$\ldots - 5'$

Fig. 7.6 A section of the macronuclear gene encoding βTP in *S. histriomuscorum*. This section contains a postfix of MDS 2 and a prefix of MDS 3, with one of the pointers remaining, while the other plus the IES were spliced out. Vertical bars indicate subsequences.

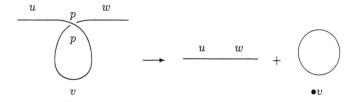

Fig. 7.7 String circular recombination I.

- *String circular recombination I:* Given a DNA molecule in which a pointer p has two occurrences. The molecule forms a hairpin so that the pointer p aligns with its copy, and the hairpin is spliced out (Fig. 7.7).
- *String circular recombination II:* This is the inverse of the first operation. Two DNA molecules, one linear and one circular and each containing one occurrence of a pointer p, become aligned in such a way that a linear molecule results in which the former circular molecule is flanked by the two occurrences of the pointer p.
- *String parallel recombination:* Two linear DNA molecules, each of which contains one occurrence of a pointer p, are aligned in a way that all data beyond the pointer p are exchanged between the two molecules (Fig. 7.8).

The second model, developed by G. Rozenberg and coworkers (2001-2006), is intramolecular and comprises the following three operations:

- *Loop recombination:* Given a DNA molecule in which two occurrences of the same pointer p flank one IES. The molecule folds so that the pointer p finds its second occurrence, and the IES is spliced out (Fig. 7.9).
- *Hairpin recombination:* Given a DNA molecule in which a pointer p has two occurrences in such a way that one is inverted. The molecule folds into a hairpin loop so that the pointer p aligns to its inverted copy, and crosses over at the hairpin crossing reinverting the inverted pointer (Fig. 7.10).
- *Double loop recombination:* Given a DNA molecule in which two occurrences of pointers p and q alternate. The molecule folds into a double loop

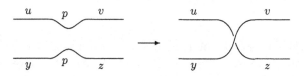

Fig. 7.8 String parallel recombination.

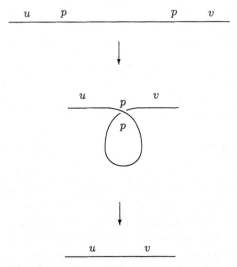

Fig. 7.9 Loop recombination.

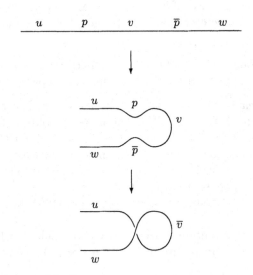

Fig. 7.10 Hairpin recombination.

in such a way that each pointer aligns to its second occurrence. In this way, the pointers are recombined so that the loop's crossing is reversed (Fig. 7.11).

In the following we provide a mathematical formalization of gene assembly in ciliates based on signed double occurrence strings and signed graphs.

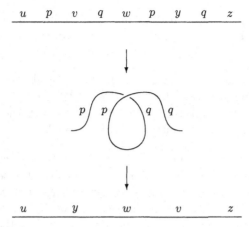

Fig. 7.11 Double loop recombination.

7.1.3 Intramolecular String Model

Let Σ be an alphabet. The *complementary alphabet* of Σ is defined as $\overline{\Sigma} = \{\overline{a} \mid a \in \Sigma\}$, where $\Sigma \cap \overline{\Sigma} = \emptyset$. The set $\Sigma \cup \overline{\Sigma}$ is termed a *signed alphabet*. Let $\Sigma^{\star} = (\Sigma \cup \overline{\Sigma})^{\star}$ denote the free monoid over the signed alphabet $\Sigma \cup \overline{\Sigma}$. Each string in $\Sigma^{\star}$ is called a *signed (double occurrence) string* over Σ.

For each symbol $a \in \Sigma$, put $\overline{\overline{a}} = a$. Thus, the mapping $\phi : \Sigma \to \overline{\Sigma} : a \mapsto \overline{a}$ is an involution and extends to an anti-morphic involution $\phi : \Sigma^{\star} \to \Sigma^{\star} :$ $x \mapsto \overline{x}$. Thus for each signed string $x = a_1 \ldots a_n$ over Σ,

$$\overline{x} = \overline{a}_n \ldots \overline{a}_1 . \tag{7.1}$$

Define the mapping $\| \cdot \| : \Sigma \cup \overline{\Sigma} \to \Sigma$ by putting $\|a\| = a = \|\overline{a}\|$.

Let $x \in \Sigma^{\star}$ be a signed string over Σ. We say that a symbol $a \in \Sigma \cup \overline{\Sigma}$ *occurs* in x, if a or $\overline{a}$ is a substring of x. Let the *domain* of the string x, $\mathrm{dom}(x)$, denote the set of all (unsigned) symbols $a \in \Sigma$ that occur in x.

A string $x \in \Sigma^{\star}$ is called *legal* if every symbol $a \in \mathrm{dom}(x)$ occurs exactly twice in x. In particular, if a legal string x contains both a and $\overline{a}$, then a is called *positive* in x. Otherwise, a or $\overline{a}$ is termed *negative* in x. Legal circular strings are similarly defined.

Example 7.1. The signed string $x = 23\overline{2}453\overline{4}5$ is legal over $\{2, 3, 4, 5\}$. While 2 and 4 are positive in x, 3 and 5 are negative in x. The signed string $23\overline{4}53\overline{4}5$ is not legal, for it has only one occurrence of the symbol 2. $\diamond$

Let $x = a_1 \ldots a_n \in \Sigma^{\star}$ be a legal string. For each symbol $a \in \mathrm{dom}(x)$, there are indices i and j with $1 \le i < j \le n$ so that $\|a_i\| = a = \|a_j\|$. The substring $x^{(a)} = a_i \ldots a_j$ is called the *a-interval* of a in x. Two symbols a

and b of dom(x) are said to *overlap* in x if the a-interval overlaps with the b-interval (i.e., if $x^{(a)} = a_i \ldots a_j$ and $x^{(b)} = a_k \ldots a_l$ then either $i < k < j < l$ or $k < i < l < j$).

Example 7.2. The signed string $x = 23\overline{2}453\overline{4}5$ is legal over $\{2, 3, 4, 5\}$. The corresponding intervals are $x^{(2)} = 23\overline{2}$, $x^{(3)} = 3\overline{2}453$, $x^{(4)} = 453\overline{4}$, and $x^{(5)} = 53\overline{4}5$. Thus, the interval of the symbol 3 overlaps with the intervals of the symbols 2, 4, and 5, and the intervals of the symbols 4 and 5 overlap. $\Diamond$

Each micronuclear gene in the gene assembly process can be described by a signed (double occurrence) string which provides the sequence and the orientation of the MDSs. For this, consider the alphabet $\Sigma_k = \{2, \ldots, k\}$, $k \geq 2$, and put $\Pi_k = \Sigma_k \cup \overline{\Sigma}_k$. The elements of Π_k are termed *pointers*. The MDS structure of a gene can be solely represented by the sequence of its pointers. For this, each MDS M_i is represented as the string $i(i+1)$, and the inverse MDS $\overline{M}_i$ is represented by the string $\overline{i+1}\,\overline{i}$, $1 < i < k$. Moreover, the first MDS M_1 and the last MDS M_k are represented in a different manner: M_1 $(\overline{M}_1)$ is described by the symbol 2 $(\overline{2})$, while M_k $(\overline{M}_k)$ is displayed by the character k $(\overline{k})$.

Example 7.3. The actin I gene in *S. nova* possesses the MDS-IES descriptor (Fig. 7.3)

$$M_3 I_1 M_4 I_2 M_6 I_3 M_5 I_4 M_7 I_5 M_9 I_6 \overline{M}_2 I_7 M_1 I_8 M_8 \ .$$

The associated MDS sequence corresponds to the legal string

$$3\,4\,4\,5\,6\,7\,5\,6\,7\,8\,9\,\overline{3}\,\overline{2}\,2\,8\,9 \ .$$

$\Diamond$

The intramolecular recombination operations for gene assembly in ciliates proposed by G. Rozenberg and coworkers can be formalized by legal strings through the following string rewriting rules:

- The *string negative rule* N_p for a pointer $p \in \Pi_k$ applies to a legal string $x = uppv$ over Σ_k,

$$N_p(uppv) = uv, \quad u, v \in \Sigma_k^\star \ . \tag{7.2}$$

Let $N = \{N_p \mid p \in \Pi_k, k \geq 2\}$ be the set of all string negative rules on signed strings.

- The *string positive rule* P_p for a pointer $p \in \Pi_k$ operates on a legal string $x = upv\overline{p}w$ over Σ_k,

$$P_p(upv\overline{p}w) = u\overline{v}w, \quad u, v, w \in \Sigma_k^\star \ . \tag{7.3}$$

Let $P = \{P_p \mid p \in \Pi_k, k \geq 2\}$ be the set of all string positive rules on signed strings.

- The *string double rule* $\mathrm{D}_{p,q}$ for pointers $p, q \in \Pi_k$ resorts to a legal string $x = upvqwpyqz$ over Σ_k,

$$\mathrm{D}_{p,q}(upvqwpyqz) = uywvz, \quad u, v, w, y, z \in \Sigma_k^\star . \qquad (7.4)$$

Let $\mathrm{D} = \{\mathrm{D}_{p,q} \mid p, q \in \Pi_k, k \geq 2\}$ be the set of all string double rules on signed strings.

A composition $\pi = \pi_n \circ \ldots \circ \pi_1$ of operations from the set $\mathrm{N} \cup \mathrm{P} \cup \mathrm{D}$ is called a *string reduction* of a string x, if π is applicable to x. A string reduction π of x is termed *successful* if the reduced string $\pi(x)$ is the empty string. The empty string stands for the abstraction of the completion of the gene assembly process for legal strings.

Example 7.4. Consider the legal string $x = 2\,3\,\overline{2}\,4\,5\,3\,\overline{4}\,5$ over Σ_5. Two successful string reductions of x are

$$(\mathrm{P}_{\overline{5}} \circ \mathrm{N}_{\overline{3}} \circ \mathrm{P}_4 \circ \mathrm{P}_2)(x) = (\mathrm{P}_{\overline{5}} \circ \mathrm{N}_{\overline{3}} \circ \mathrm{P}_4)(\overline{3}\,4\,5\,3\,\overline{4}\,5) = (\mathrm{P}_{\overline{5}} \circ \mathrm{N}_{\overline{3}})(\overline{3}\,\overline{3}\,\overline{5}\,5)$$
$$= \mathrm{P}_{\overline{5}}(\overline{5}\,5) = \epsilon \,,$$

and

$$(\mathrm{P}_2 \circ \mathrm{P}_{\overline{4}} \circ \mathrm{D}_{3,5})(x) = (\mathrm{P}_2 \circ \mathrm{P}_{\overline{4}})(2\,\overline{4}\,\overline{2}\,4) = \mathrm{P}_2(2\,\overline{2}) = \epsilon \,.$$

$$\diamond$$

Theorem 7.5. *Let $k \geq 2$ be an integer. Each legal string over Σ_k possesses a successful string reduction.*

Proof. The proof uses induction over k. Let x be a legal string over Σ_2. Clearly, the string x can be successfully reduced by the positive and negative rules.

Let x be a legal string over Σ_k, $k > 2$. Let κ denote k or $\overline{k}$. Consider three cases:

- If x has the form $u\kappa\kappa v$, then $\mathrm{N}_\kappa(x) = uv$, and by induction uv has a successful string reduction.
- If x has the shape $u\kappa v\overline{\kappa}w$ then $\mathrm{P}_\kappa(x) = u\overline{v}w$, and by induction $u\overline{v}w$ is successfully reducible.
- Otherwise, x is of the form $u\kappa v\kappa w$ so that v is not empty. First, suppose that the k-interval of x does not overlap with another interval of x. Then v is a legal string over Σ_{k-1} and thus by induction is successfully reducible. This provides the string $u\kappa\kappa w$ which is reducible by the string negative rule to the string uw. By induction, the string uw has a successful string reduction and hence x is successfully reducible.
  Second, suppose that there is a symbol $l \in \Sigma_{k-1}$ so that the k-interval of x overlaps with the l-interval of x. If l is negative in x, then the string

double rule yields the string $y = D_{\kappa,l}(x)$ in which κ and l are deleted. Thus by induction, the string y is successfully reducible.

This leaves us with the case that the k-interval of x overlaps only with l-intervals of x, where $l \in \Sigma_{k-1}$ is positive in x. For each such symbol l, the string positive rule P_l is applied in order to eliminate both l and $\bar{l}$. These elimination steps are repeatedly carried out resulting in a string x' of the form $y\kappa\kappa z$. This string can be subject to the negative rule yielding the string $N_\kappa(x') = yz$. By induction, the string yz possesses a successful string reduction and thus x is successfully reducible, as required.   □

7.1.4 Intramolecular Graph Model

A *signed graph* G is a triple (V, E, μ), where (V, E) is an undirected graph and $\mu : V \to \{\pm 1\}$ is a vertex labelling. A vertex v in G is called *positive* if $\mu(v) = +1$, and *negative* if $\mu(v) = -1$. Put $V^+ = \{v \in V \mid \mu(v) = +1\}$ and $V^- = V \setminus V^+$. Let G^+ be the signed subgraph of G induced by V^+, and let G^- be the signed subgraph of G induced by V^-. A signed graph is *all-positive* if $V = V^+$, and *all-negative* if $V = V^-$.

Each legal string x over Σ is associated with a signed graph $G_x = (V_x, E_x, \mu_x)$ by the following settings:

- $V_x = \mathrm{dom}(x)$,
- $E_x = \{ab \mid a \text{ and } b \text{ overlap in } x\}$, and
- $\mu_x(a) = +1$ if a is positive in x, and $\mu_x(a) = -1$ if a is negative in x.

Each micronuclear gene in the gene assembly process can be described by a signed graph which provides the overlap of the pointers.

Example 7.6. The signed graph for the micronuclear gene actin I in *S. nova* is given in Fig. 7.12.   ◇

Let $G = (V, E, \mu)$ be a signed graph and let $U \subseteq V$. If the subgraph of G induced by U is replaced by its complementary graph inclusively complementing the signs of the vertices in U, the resulting signed graph $\mathrm{com}_U(G) = (V, E', \mu')$ is said to be the *U-complement* of G. In particular, if U is given by the *neighborhood* $N_G(v) = \{u \mid uv \in E\} \cup \{v\}$ of a vertex v in

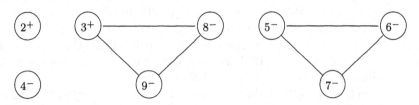

Fig. 7.12 Signed graph of the micronuclear gene actin I in *S. nova* (Fig. 7.3).

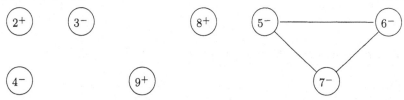

Fig. 7.13 Local complement $\text{loc}_3(G)$ of the signed graph G for the micronuclear gene actin I in *S. nova* (Fig. 7.12).

G, then the U-complement of G is called the *local complement* of G at v and is denoted by $\text{loc}_v(G)$ (Fig. 7.13). For each vertex v in G, let $G - v$ denote the subgraph of G that is induced by the set $V \setminus \{v\}$.

The intramolecular recombination operations for gene assembly in ciliates proposed by G. Rozenberg and coworkers can be formalized by signed graphs as follows:

- The *graph negative rule* for a vertex p applies to G if p belongs to V^- and is isolated. The resulting graph $N_p(G)$ is the signed graph $G - p$. Let $N = \{N_p \mid p \geq 2\}$ be the set of all graph negative rules on signed graphs.
- The *graph positive rule* for a vertex p applies to G if p belongs to V^+. The resulting graph $P_p(G)$ is the signed graph $\text{loc}_p(G) - p$. Let $P = \{P_p \mid p \geq 2\}$ be the set of all graph positive rules on signed graphs.
- The *graph double rule* for two vertices p and q applies to G if p and q both belong to V^- and are adjacent. The resulting graph $D_{p,q}(G) = (V \setminus \{p, q\}, E', \mu')$ is obtained from G so that μ' equals μ restricted to $V \setminus \{p, q\}$ and E' is derived from E by complementing the edges that join vertices in $N_G(p)$ with vertices in $N_G(q)$. Let $D = \{D_{p,q} \mid p, q \geq 2\}$ be the set of all graph double rules on signed graphs.

Example 7.7. Consider the legal string $x = 2\,3\,\overline{2}\,4\,5\,3\,\overline{4}\,5$ over Σ_5. The associated signed graph G_x is shown in Fig 7.14. Moreover, the string $P_4(x) = 2\,3\,\overline{2}\,\overline{3}\,5\,5$ corresponds to the signed graph $P_4(G_x)$ in Fig 7.15, and the string $D_{3,5}(x) = 2\,\overline{4}\,\overline{2}\,4$ is associated with the signed graph $D_{3,5}(G_x)$ in Fig 7.16. ◇

Let G be a signed graph. A composition $\pi = \pi_1 \circ \ldots \circ \pi_n$ of operations from the set $N \cup P \cup D$ is a *successful strategy* for the graph G if the reduced signed graph $\pi(G)$ is the *empty graph* (i.e., the graph with empty vertex

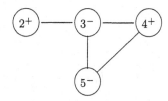

Fig. 7.14 Signed graph for the legal string $x = 2\,3\,\overline{2}\,4\,5\,3\,\overline{4}\,5$.

Fig. 7.15 Signed graph for the string $P_4(x) = 2\,3\,\overline{2}\,3\,\overline{5}\,5$.

Fig. 7.16 Signed graph for the string $D_{3,5}(x) = 2\,\overline{4}\,\overline{2}\,4$.

set). The empty graph stands for the abstraction of the completion of the gene assembly process for signed graphs. In view of the analogy between string reduction rules and graph reduction rules, Theorem 7.5 can be stated in terms of signed graphs.

Theorem 7.8. *Let $k \geq 2$ be an integer. The signed graph associated with a legal string over Σ_k has a successful strategy.*

String and graph reductions suggest that gene assembly is a sequential process. However, parallelism is a natural phenomenon in biomolecular processes. Therefore, gene assembly should be considered as a parallel process. Intuitively, a set of operations can be applied in parallel to a gene pattern if and only if the operations can be sequentially applied to the pattern in any order. This view is consistent with the notion of concurrency in Computer Science.

Let R be a finite subset of rules from the set $\mathsf{N} \cup \mathsf{P} \cup \mathsf{D}$, and let G be a signed graph. The rules in R are *applicable in parallel* to G if for any ordering of the operations $\pi_1, \ldots, \pi_n$ in R, the composition $\pi = \pi_n \circ \ldots \circ \pi_1$ is applicable to G. In particular, two operations π and τ are applicable in parallel to G if both $\pi \circ \tau$ and $\tau \circ \pi$ are applicable to G. Notice that if the rules in R are applicable in parallel to a signed graph G, then by definition the rules in any subset of R are applicable in parallel to G. However, the converse it not true as demonstrated by the following:

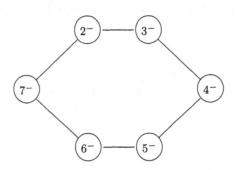

Fig. 7.17 Signed graph G.

Example 7.9. Consider the signed graph G in Fig. 7.17. It can be easily checked whether any two rules from the set $R = \{D_{2,3}, D_{4,5}, D_{6,7}\}$ are applicable in parallel to G. However, the rules in R are not applicable in parallel to G. To see this, observe that the signed graph $(D_{2,3} \circ D_{4,5})(G)$ is the isolated all-negative graph on the vertex set $\{6,7\}$, and thus $D_{6,7}$ is not applicable to this graph. $\diamond$

The main result on string or graph reduction in ciliates states that reduction is the same, regardless of the sequential order in which the rules are applied. To prove this, we need two basic assertions.

Lemma 7.10. *Let G be a signed graph with vertices p and q. If $D_{p,q}$ is applicable to G, then $D_{p,q}(G) = P_{\overline{p}}(P_{\overline{q}}(\mathrm{loc}_p(G)))$.*

Proof. Let $x = upvqwpyqz$ be a legal string. Thus $D_{p,q}$ is applicable to x and hence $D_{p,q}(x) = uywvz$. On the other hand,

$$P_{\overline{p}}(P_{\overline{q}}(\mathrm{loc}_p(x))) = P_{\overline{p}}(P_{\overline{q}}(u\overline{pvqw}pyqz)) = P_{\overline{p}}(P_{\overline{q}}(u\,\overline{p}\,\overline{w}\,\overline{q}\,\overline{v}\,\overline{p}\,y\,q\,z))$$

$$= P_{\overline{p}}(u\,\overline{p}\,\overline{w}\,\overline{\overline{v}}\,\overline{p}\,y\,z) = P_{\overline{p}}(u\,\overline{p}\,\overline{w}\,\overline{y}\,p\,v\,z)$$

$$= u\,\overline{\overline{w}}\,\overline{\overline{y}}\,v\,z = uywvz\,.$$

$\square$

Lemma 7.11. *If $G = (V, E, \mu)$ is a signed graph and $U_1, U_2 \subseteq V$, then $\mathrm{com}_{U_1}(\mathrm{com}_{U_2}(G)) = \mathrm{com}_{U_2}(\mathrm{com}_{U_1}(G))$.*

Proof. Let $G_1 = (V_1, E_1, \mu_1)$ and $G_2 = (V_2, E_2, \mu_2)$ be graphs so that $G_1 = \mathrm{com}_{U_1}(\mathrm{com}_{U_2}(G))$ and $G_2 = \mathrm{com}_{U_2}(\mathrm{com}_{U_1}(G))$. Let $p \in V$. If $p \notin U_1 \cup U_2$, then the neighborhood of p and the sign of p are the same in G, G_1, and G_2. If $p \in U_1 \setminus U_2$, then the sign of p changes but is the same in both G_1 and G_2. Furthermore, the neighborhood of p is complemented in both G_1 and G_2. Finally, if $p \in U_1 \cap U_2$, then the sign of p changes twice and is the same in G, G_1 and G_2. Moreover, the neighborhood of p is complemented twice and is therefore the same in G, G_1 and G_2. It follows that the signed graphs G_1 and G_2 coincide. $\square$

Let us begin with a special case of the main result.

Lemma 7.12. *Let $G = (V, E, \mu)$ be a signed graph. If $\pi, \tau \in \mathrm{N} \cup \mathrm{P} \cup \mathrm{D}$ are applicable in parallel to G, then $\pi(\tau(G)) = \tau(\pi(G))$.*

Proof. If $\pi \in \mathrm{N}$ or $\tau \in \mathrm{N}$, then the result is clear. Otherwise, let $\pi, \tau \in \mathrm{P} \cup \mathrm{D}$. Each operation in P is a composition of a local complementation and a vertex removal. Moreover, by Lemma 7.10, each operation in D can be represented by a composition of local complementations and vertex removals. But by Lemma 7.11, local complementations commute. Furthermore, vertex removals commute (i.e., for any $p, q \in V$, $(G - p) - q = (G - q) - p$), and local complementations and vertex removals commute, that is, for any $U \subseteq V$ and $p \in V$, $\mathrm{com}_U(G) - p = \mathrm{com}_U(G - p)$. Hence, by commutativity, the result follows.   □

These preliminary assertions lead us to the main result.

Theorem 7.13. *Let G be a signed graph and let $R \subseteq N \cup P \cup D$ be a set of rules applicable in parallel to G. For any two compositions π and π' of the rules in R, $\pi(G) = \pi'(G)$.*

Proof. There exists a sequence $\pi = \pi_0, \pi_1, \ldots, \pi_n = \pi'$ of permutations of π so that $\pi_i = \pi_{i,2} \circ \alpha_i \circ \beta_i \circ \pi_{i,1}$ and $\pi_{i+1} = \pi_{i,2} \circ \beta_i \circ \alpha_i \circ \pi_{i,1}$, where $\pi_{i,1}$ and $\pi_{i,2}$ are compositions and α_i and β_i are rules in R. Thus, it suffices to consider compositions of the form $\pi = \pi_2 \circ \alpha \circ \beta \circ \pi_1$ and $\pi' = \pi_2 \circ \beta \circ \alpha \circ \pi_1$, where π_1 and π_2 are compositions and α and β are rules in R. Clearly, $\pi(G) = \pi'(G)$ if and only if $(\alpha \circ \beta)(\pi_1(G)) = (\beta \circ \alpha)(\pi_1(G))$. But the latter follows directly from Lemma 7.12.   □

7.1.5 Intermolecular String Model

The intermolecular model for gene assembly in ciliates postulated by L. Kari and L. Landweber (1998) can be formalized in the framework of legal strings by the following string rewriting rules:

- The *string circular rule I* for a pointer $p \in \Pi_k$ applies to a legal string $x = upvpw$ over Σ_k,

$$C_p^{(1)}(upvpw) = uw + \bullet v, \quad u, v, w \in \Sigma_k^\star. \tag{7.5}$$

Let $C^{(1)} = \{C_p^{(1)} \mid p \in \Pi_k, k \geq 2\}$ be the set of all string circular rules I on legal strings.

- The *string circular rule II* for a pointer $p \in \Pi_k$ operates on legal strings $x = upv$ and $x' = \bullet pw$ over Σ_k,

$$C_p^{(2)}(upv + \bullet pw) = uwv, \quad u, v, w \in \Sigma_k^\star. \tag{7.6}$$

Let $C^{(2)} = \{C_p^{(2)} \mid p \in \Pi_k, k \geq 2\}$ be the set of all string circular rules II on legal strings.

- The *string parallel rule* for a pointer $p \in \Pi_k$ resorts to legal strings $x = upv$ and $x' = ypz$ over Σ_k,

$$L_p(upv + ypz) = uz + yv, \quad u, v, y, z \in \Sigma_k^\star. \tag{7.7}$$

Let $L = \{L_p \mid p \in \Pi_k, k \geq 2\}$ be the set of all string parallel rules on legal strings.

These rules usually operate on multisets of legal strings so that each string has a *multiplicity* as in the definition of splicing systems.

Example 7.14. We have $C_3^{(1)}(23\overline{2}453\overline{4}5) = 2\overline{4}5 + \bullet\overline{2}45$, $C_3^{(2)}(23\overline{2}4 + \bullet53\overline{4}5) = 2\overline{4}55\overline{2}4$, and $L_3(23\overline{2}4 + 53\overline{4}5) = 2\overline{4}5 + 5\overline{2}4$. $\Diamond$

A composition $\pi = \pi_n \circ \ldots \circ \pi_1$ of operations from $C^{(1)} \cup C^{(2)} \cup L$ is a *string reduction* of a string x, if π is applicable to x. A string reduction π of x is *successful* if the reduced string $\pi(x)$ is the empty string (more generally, a multiset of empty strings). The empty string stands for the abstraction of the completion of the gene assembly process for legal strings.

This model cannot deal with legal strings in which a pointer is inverted. Therefore, we make two assumptions: Each legal string is available in two copies and the inversion of each legal string is available as well. While the first assumption is vital in the intermolecular model, the second is quite natural whenever double-stranded DNA molecules are modeled.

Theorem 7.15. *Let $k \geq 2$ be an integer. Each legal string over Σ_k possesses a successful string reduction in the intermolecular model.*

Proof. Let x be a legal string over Σ_k. We consider three cases:

- If $x = uppv$ for some pointer $p \in \Pi_k$, then

$$C_p^{(1)}(uppv) = uv + \bullet\epsilon = N_p(uppv) + \bullet\epsilon.$$

- If $x = upv\overline{p}w$ for some pointer $p \in \Pi_k$, then by assumption the inverted legal string is also available and thus we obtain

$$\begin{aligned}
(L_{\overline{p}} \circ L_p)(u\,p\,v\,\overline{p}\,w + \overline{u\,p\,v\,\overline{p}\,w}) &= (L_{\overline{p}} \circ L_p)(u\,p\,v\,\overline{p}\,w + \overline{w}\,p\,\overline{v}\,\overline{p}\,\overline{u}) \\
&= L_{\overline{p}}(u\,\overline{v}\,\overline{p}\,\overline{u} + \overline{w}\,v\,\overline{p}\,w) \\
&= u\,\overline{v}\,w + \overline{w}\,v\,\overline{u} \\
&= P_p(u\,p\,v\,\overline{p}\,w) + P_p(\overline{w}\,p\,\overline{v}\,\overline{p}\,\overline{u}) \\
&= P_p(u\,p\,v\,\overline{p}\,w) + P_p(\overline{u\,p\,v\,\overline{p}\,w}) \\
&= P_p(u\,p\,v\,\overline{p}\,w + \overline{u\,p\,v\,\overline{p}\,w}).
\end{aligned}$$

- If $x = upvqwpyqz$ for some pointers $p, q \in \Pi_k$, then

$$\begin{aligned}
(C_q^{(2)} \circ C_p^{(1)})(upvqwpyqz) &= C_q^{(2)}(uyqz + \bullet vqw) \\
&= C_q^{(2)}(uyqz + \bullet qwv) \\
&= uywvz \\
&= D_{p,q}(upvqwpyqz).
\end{aligned}$$

These equations show that each string reduction of x in the intermolecular model can be simulated by a string reduction of x in the intramolecular model, and vice versa. Hence, by Theorem 7.8, the result follows. $\square$

7.2 Biomolecular Computing

A diverse library of easy-to-use biological components that expands the capabilities to probe and control cell behavior is under construction.

7.2.1 Gene Therapy

Many diseases result from defective genes, which are often caused by genetic mutations. Such mutations can be inherited or induced by several stress factors. Mutations can persist and can be passed down through generations. Defective alleles of a gene often have no harmful effects since our genomes are diploid, that is, two copies of nearly all genes are available. The only exception to this rule are the genes found on the male sex chromosomes. In the majority of situations, one normal gene is sufficient to avoid all symptoms of disease. However, a disease phenotype will develop if the mutation persists through the two alleles. As a consequence, the gene cannot be expressed into a protein or the expressed protein has an altered 3D-fold. In either case, a crucial physiological activity of the cell might be depleted. In this context, a good strategy to treat diseases based on a mutational gene pattern is either "correcting" the aberrant gene or "supplementing" the cell with a copy of the healthy (wild-type) gene. The cells can then produce the correct protein and consequently eliminate the root causative of the disease. Gene therapy can target somatic (body) or germ (egg and sperm) cells. Somatic gene therapy changes the genome of the recipient, but this change is not passed along to the next generation. In contrast, germline therapy alters the egg and sperm cells and these changes are passed to the offspring. Today, somatic gene therapy is primarily at the experimental stage, while germline therapy is the subject of much debate.

In most gene therapy studies, a healthy gene is inserted into the genome to replace an aberrant gene. The most common delivery system is vector mediated. A *vector* is a carrier molecule delivering a healthy gene into the recipient's target cells. The most commonly used vectors are viruses that are genetically prepared to carry normal human DNA, since viruses have found a way of encapsulating and delivering their genes to human cells in a pathogenic manner. The most common viruses for gene therapy are retroviruses and adenoviruses. The former can create double-stranded DNA copies of their RNA genome and integrate these copies into the chromosomes of the host cells, while the latter have a linear double-stranded DNA genome capable of replicating in the nucleus of mammalian cells using the host's biosynthesis machinery. Besides vector-based gene supply, there are several other options for gene delivery. One method is to directly introduce the therapeutic DNA product into the target cells. However, this approach only works with certain tissues and requires large amounts of DNA. Another method involves creating an artificial lipid sphere with an aqueous core containing the therapeutic DNA. This liposome is able to pass the DNA through the target cell's membrane.

Gene therapy faces many obstacles before it can be considered an effective approach for treatment. First, the healthy DNA introduced into target cells must remain functional and the cells containing the healthy DNA must be long-lived and stable. Problems with integrating DNA into the genome and the rapidly dividing nature of many cells prevent gene therapy from achieving long-term benefits. Thus, patients will have to undergo multiple rounds of gene therapy. Second, the immune system is stimulated each time a foreign object is introduced into human tissue, and this stimulus may reduce the effectiveness of the gene therapy. Moreover, the immune system's enhanced response to invaders previously encountered hampers a repeated treatment. Third, viruses may cause potential problems for the recipient, like toxicity, immune and inflammatory responses, and gene control and targeting issues. Furthermore, the viral vector may recover its ability to cause disease inside the recipient. Finally, some of the most commonly occurring disorders, such as heart disease, high blood pressure, diabetes, and arthritis, are caused by combined defects in various genes. These multigene disorders are difficult to treat effectively using gene therapy.

7.2.2 Anti-Sense Technology

Anti-sense technology is based on a widespread mechanism in natural gene expression control. Anti-sense technology uses anti-sense sequences that specifically pair and subsequently inhibit target sense sequences. Anti-sense sequences are DNA or RNA or chemically modified nucleotide sequences, while the usual target in anti-sense strategies is mRNA. Anti-sense sequences bind to the complementary part of the single-stranded mRNA and block its translation. The double-stranded hybrid is further targeted for degradation. Anti-sense-mediated steric hindrance can in addition to translation affect RNA processing, RNA transport, and transcription (in the case of viral RNA). RNA level intervention by anti-sense offers several opportunities for gene control. Indeed, genomic diversity expands at the RNA level through RNA processing, and differences in processing patterns can be manipulated through anti-sense intervention. The first anti-sense drug that achieved market clearance was vitravene, treating a condition called cytomegalovirus retinitis in people with AIDS.

Gene Expression Anti-Switches

Anti-sense technology can be used to regulate gene expression in a ligand-dependent manner. This was demonstrated by T. Bayer and C. Smolke (2005), providing an anti-switch that can be used for programming cellular behavior and genetic networks with respect to cellular state and environmental stimuli. For this, a trans-acting RNA-based regulator termed *anti-switch* was designed that regulates target expression in response to a ligand. The anti-switch uses an anti-sense domain to control gene expression and an aptamer domain to recognize specific effector ligands. Aptamers are DNA species that bind specific ligands and are generated through in vitro selection or systematic evolution of ligands by exponential enrichment (SELEX). Ligand binding at the aptamer domain induces a conformational change that allows the anti-sense domain (15 nt) to interact with a target mRNA to affect translation. In the absense of the ligand, the anti-sense domain is sequestered in an anti-sense stem preventing target binding (Fig. 7.18). The anti-switch was tested in *Saccharomyces cerevisiae* (baker's yeast) using theophylline as a ligand.

RNA Interference

Another useful anti-sense technology is RNA-mediated interference (RNAi), which provides a pathway conserved in most eukaryotic organisms as a form of innate immunity against viruses and other foreign gene material. The RNAi pathway is initiated by either exogenous or endogenous double-stranded RNA that is cut by the ribonuclease dicer. This enzyme binds and cleaves longer double-stranded RNA into fragments of 19 to 25 bp. These short double-stranded fragments, termed small interfering RNA (siRNA), are separated into single-stranded RNA and integrated into an RNA-induced silencing com-

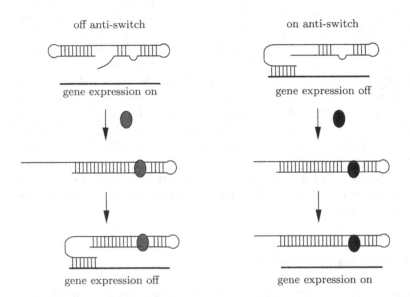

Fig. 7.18 Schematic behavior of anti-switch regulator.

plex (RISC). After integration into RISC, siRNA can anneal to complementary stretches of the target mRNA and, as mentioned before, mediate the silencing of the gene expression.

In mammalian cells, it is known that double-stranded RNA of 30 bp or longer can trigger interferon responses, which assist the immune system by inhibiting viral replication. These responses are intrinsically sequence-non-specific. Hence, as an experimental and therapeutic agent, the application of RNA interference is very limited.

RNAi was recently used by Y. Benenson and coworkers (2007) to implement Boolean logic in human kidney cells. For this, Boolean gates were constructed whose input is comprised of two or more mRNA species. These mRNAs encode the same protein, but have different non-coding regions. The corresponding protein is the circuit's output. The logic values "true" and "false" are represented by the concentrations of the molecules. An OR gate can be designed in a straightforward manner. If at least one mRNA species is present, the output protein will be produced.

An AND gate makes use of molecular mediators in the form of siRNAs, since they are known to target untranslated regions. For this, selective inhibitory links must be designed between the endogenous inputs and these siRNAs. If an input is present, it blocks the linked siRNA. Thus, if all inputs are present at the same time, all siRNAs are blocked and the output is generated from the mRNA.

A NOT gate uses a selective activating link between the endogenous input and siRNA. If an input is present the linked siRNA is induced, while if an input is absent the associated siRNA is inhibited. Currently, the evaluator modules were implemented, but not the sensor modules providing the link between input and siRNA.

7.3 Cell-Based Finite State Automata

The first autonomous finite state machine working in a living cell was proposed by Y. Sakakibara and coworkers (2006). This approach is based on the length-encoding automaton model (Sect. 6.2.2) and was tested in *E. coli* cells.

Data Representation

Let $M = (\Sigma, S, \delta, s_0, F)$ be a finite automaton with state set $S = \{s_0, \ldots, s_m\}$.

Similar to the length-encoding automaton model, each symbol a in Σ is encoded by a single RNA strand $\sigma(a)$. In this way, each input string

$x = a_1 \ldots a_n$ over Σ will be encoded by a single RNA strand consisting of alternating symbol encodings and spacers,

$$\sigma(x) = 5' - X_1 \ldots X_m \sigma(a_1) X_1 \ldots X_m \sigma(a_2) \ldots X_1 \ldots X_m \sigma(a_n) - 3' , \quad (7.8)$$

where $X_1, \ldots, X_m$ is the *spacer* sequence. A start codon should be present in front of the 5' end for translation purposes. Moreover, an RNA sequence should be appended at the 3' end of the input strand containing, in turn, one of the stop codons, UAA, UAG or UGA, and the mRNA sequence of a gene of interest, that is,

$$5' - R_1 \ldots R_u UAAY_1 \ldots Y_v Z_1 \ldots Z_w - 3' , \quad (7.9)$$

where $R_1 \ldots R_u$ and $Y_1 \ldots Y_v$ are RNA subsequences flanking a stop codon (UAA), and $Z_1 \ldots Z_w$ is the mRNA strand of a gene.

Similar to the length-encoding automaton model, each transition rule $\delta(s_i, a) = s_j$, $a \in \Sigma$, is encoded by the single strand

$$3' - \overline{X}_{i+1} \ldots \overline{X}_m \overline{\sigma(a)} \overline{X}_1 \ldots \overline{X}_j - 5' , \quad (7.10)$$

where $\overline{X}$ denotes the Watson-Crick complement of the nucleotide X, and $\overline{\sigma(a)}$ refers to the Watson-Crick complement of the RNA sequence $\sigma(a)$. In this model, however, each transition rule is implemented by a sequence of tRNA molecules so that the concatenation of the corresponding anti-codons equals as RNA molecule the single strand (7.10). To this end, artificial tRNA molecules with four- and five-base anti-codons are employed.

Example 7.16. Consider the finite state automaton M in Fig. 6.23. Put $\sigma(a) = $ AGGU, $\sigma(b) = $ GCGC and take as spacer the nucleotide A. The word $x = abab$ is then encoded as $5' - $ AAGGUAGCGCAAGGUAGCGC $ - 3'$. The encoding of the state transitions is illustrated in Table 7.1, while the implementation of the state transitions by tRNA molecules is shown in Table 7.2. The transition rule $\delta(s_0, a) = s_1$ is encoded by the concatenation of the anti-codons corresponding to the tRNA molecules T_1 and T_2,

$$3' - \underbrace{UUC}\ \underbrace{CAU} - 5' .$$

The remaining transition rules are encoded by singleton tRNA molecules. $\Diamond$

Computation

An input string is encoded by an mRNA molecule consisting of the concatenated strands in (7.8) and (7.9). The computation of this mRNA molecule is accomplished by the biosynthesis mechanism of the living cell.

Table 7.1 Encoding of state transitions in finite state automaton (Fig. 6.23).

Transition	Encoding
$\delta(s_0, a) = s_1$	$3' - \text{UUCCAU} - 5'$
$\delta(s_0, b) = s_0$	$3' - \text{UCGCG} - 5'$
$\delta(s_1, a) = s_0$	$3' - \text{UCCA} - 5'$
$\delta(s_1, b) = s_1$	$3' - \text{CGCGU} - 5'$

Table 7.2 Encoding of state transitions by tRNA molecules.

tRNA	Anti-Codon	Transition Rule
T_1	$3'-\text{UUC}-5'$	$\delta(s_0, a) = s_1$
T_2	$3'-\text{CAU}-5'$	
T_3	$3'-\text{UCCA}-5'$	$\delta(s_1, a) = s_0$
T_4	$3'-\text{UCGCG}-5'$	$\delta(s_0, b) = s_0$
T_5	$3'-\text{CGCGU}-5'$	$\delta(s_1, b) = s_1$

For this, input strings need to be transfected into the living cell, and if not naturally available, the tRNA molecules in the cell must be transfected, too. The computation is carried out by tRNA molecules in such a way that if the input string is accepted, the input string including the gene of interest is translated (as a single product), and if the input string is not accepted then the translation ends at the stop codon made available in the substrand (7.9). In the latter case, the gene of interest will not be translated. This requires that the appended substrand $5' - R_1 \ldots R_u \text{UAAY}_1 \ldots Y_v - 3'$ containing the stop codon must be appropriately designed.

Example 7.17. In view of the previous example, the input string $x = aaaa$ is accepted by the automaton via the transitions $s_0 \xrightarrow{a} s_1$, $s_1 \xrightarrow{a} s_0$, $s_0 \xrightarrow{a} s_1$, and $s_1 \xrightarrow{a} s_0$. The corresponding mRNA molecule is given by

$$5' - \text{AAGGUAAGGUAAGGUAAGGUAUAAGG}Z_1 \ldots Z_w - 3'$$

where AUAAGG is the appended substrand containing the stop codon UAA. The computation carried out by the tRNA molecules translates the gene of interest because the stop codon is out of phase:

$$5' - \text{AAG GUA AGGU AAG GUA AGGU AUA AGG } Z_1 \ldots Z_w - 3'$$
$$3' - \underbrace{\text{UUC}} \; \underbrace{\text{CAU}} \; \underbrace{\text{UCCA}} \; \underbrace{\text{UUC}} \; \underbrace{\text{CAU}} \; \underbrace{\text{UCCA}} \; \underbrace{\text{UAU}} \; \underbrace{\text{UCC}} \; \overline{Z}_1 \ldots \overline{Z}_w - 5' \, .$$

The input string $x = aaa$ is not accepted by the automaton and is encoded by the mRNA molecule

$$5' - \text{AAGGUAAGGUAAGGUAUAAGG}Z_1 \ldots Z_w - 3' \, .$$

The computation accomplished by the tRNA molecules ends at the stop codon UAA so that the gene of interest will not be translated:

$$5' - \text{AAG GUA AGGU AAG GUA UAA AGG } Z_1 \ldots Z_w - 3'$$
$$3' - \underbrace{\text{UUC}} \ \underbrace{\text{CAU}} \ \underbrace{\text{UCCA}} \ \underbrace{\text{UUC}} \ \underbrace{\text{CAU}}$$

Theorem 7.18. *Let M be the finite automaton in Figure 6.23. In view of the given encodings of states and transitions of M, a string is accepted by M if and only if the gene of interest is translated.*

Proof. Let $k \geq 0$ be an integer and let m be the length of the gene of interest. First, the accepted string a^{2k} has encoding length $k \cdot 10 + 6 + m$, and the sequence of tRNA molecules $(T_1 T_2 T_3)^k$ translates the first $k(3+3+4) = k \cdot 10$ bases. But the stop codon starts at the second position in the appended strand of length 6. Hence, the stop codon is not in phase and thus will not be translated.

Second, the non-accepted string a^{2k+1} has encoding length $(k \cdot 10 + 5) + 6 + m$, and the sequence of tRNA molecules $(T_1 T_2 T_3)^k T_1 T_2$ translates the first $k(3+3+4)+3+3 = k \cdot 10 + 6$ bases. Thus, the first base in the appended strand is translated by the last tRNA molecule T_2. Hence, the next translated codon is the stop codon.

Third, consider an arbitrary string containing the symbol b. This symbol has encoding length 5 and is processed by M in state s_0 or s_1 without leaving the state. The tRNA molecules corresponding to these transitions have encoding length 5, too. Therefore, the translation of the symbol b leaves the phase invariant. □

During translation, tRNA molecules with anti-codons longer than three bases (e.g., $3' - \text{UCCA} - 5'$), will eventually compete with naturally occurring tRNA molecules whose anti-codons are prefixes (e.g., $3' - \text{UCC} - 5'$), encoding the amino acid glycine. Hence, the success of a computation in this model crucially depends on the concentration of available tRNA molecules. The experiments showed that a computation by a single *E. coli* cell is not effective and accurate, while a colony of *E. coli* cells provides more reliable computations. Since bacterial cells can multiply to over a million cells overnight, these in vivo computations might offer a massive amount of parallelism.

7.4 Anti-Sense Finite State Automata

Finite state automata operating autonomously at the molecular scale can be used conceptually for applications in the living cell. This was first demonstrated by E. Shapiro and coworkers (2004), who built a small finite state automaton from DNA strands and enzymes. This automaton uses anti-sense technology to carry out molecular diagnosis and therapy.

7.4.1 Basic Model

The Shapiro automaton model can, in principle, be used to construct molecular automata that control drug release. The input of such an automaton is given by a combination of molecular indicators given by mRNA molecules at specific levels, and the output provides a drug in the form of an oligonucleotide. The automaton has two states: yes (y) and no (n). The computation starts in the state y and if it ends in that state, the result is a positive diagnosis; otherwise, it is a negative diagnosis. The diagnosis is established by a series of transitions and each transition tests for high or low levels of a particular indicator (Fig. 7.19). Once the automaton enters the state n, it remains in this state for the duration of the computation.

The computation is purely stochastic, since the transitions are sensitively controlled by the concentrations of the indicators. A present indicator increases the probability of a positive transition by the corresponding molecular indicator and decreases the probability of its competing negative transition, and vice versa if the indicator is absent.

Fine control over the diagnosis is attained by administering a biologically active molecule on a positive diagnosis, and its suppressor molecule on a negative diagnosis. Because a single molecular automaton of Shapiro type cannot perform this task, two automata are employed. One releases a drug molecule on a positive diagnosis and the other administers a drug-suppressor molecule on a negative diagnosis. The ratio between drug and drug-suppressor released by a population of automata of the two types determines the final active drug concentration.

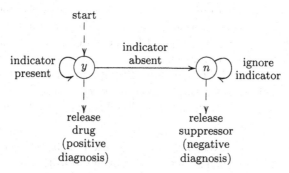

Fig. 7.19 Transition diagram of diagnostic automaton.

7.4.2 Diagnostic Rules

Diagnostic rules encode medical knowledge and are given by an if–then statement; this if–then mechanism is a new element of molecular computation.
The corresponding Boolean expression consists of a conjunction of molecular
indicators, and if this expression becomes "true" then an *anti-sense drug* is
released. This drug is designed to prevent the expression of a certain gene.
For this, the drug specifically binds to an mRNA strand of the gene and
thereby inhibits its translation into a protein.

Example 7.19. The diagnostic rule for prostate cancer states that if the genes
ppap2p and gstp1 are underexpressed and the genes pim1 and hpn are overexpressed, then the anti-sense oligonucleotide $5' - \text{GTTGGTATTGCACAT} - 3'$ for
the gene mdm2 is administered. The gene mdm2 is an important negative
regulator for the p53 tumor suppressor protein. The p53 protein (also TP53)
is a crucial transcription factor participating in the regulation of the cell
cycle. It acts as a tumor suppressor by preventing mutations in the genome.
For this, p53 directs mutated sequences for repair or induces programmed
cell death (apoptosis), thus preventing damaged DNA to propagate in the
cell.

   The diagnostic rule for small cell lung cancer states that if the genes
ascl1, gria2, insm1, and pttg are overexpressed, then the anti-sense strand
$5' - \text{TCTCCCAGCGTGCGCCAT} - 3'$ (oblimersen) is released. $\diamondsuit$

7.4.3 Diagnosis and Therapy

This kind of diagnostic rule might be implemented by the molecular automaton model described in Section 6.2.1. This model contains diagnostic
molecules that encode diagnostic rules, transition molecules that realize
automaton transitions, and hardware molecules given by the restriction
enzyme FokI.

   A diagnostic rule is encoded by a DNA molecule that consists of a diagnostic moiety and a drug-administering moiety. The diagnostic moiety encodes
the indicators of the corresponding diagnostic rule by a double-stranded DNA
molecule. The indicators correspond to the automaton's symbols, and the
automaton computes the diagnostic moiety symbol by symbol (Fig. 7.20).

   Each indicator is realized by a pair of competing transition molecules. For
instance, the indicator for the overexpressed gene $\text{pim1}^\uparrow$ gives rise to two
transition rules, which are realized by double-stranded DNA molecules. The
transition rule $y \xrightarrow{\text{pim1}^\uparrow} n$ is *initially active* that is able to interact with the
diagnostic moiety. However, if the concentration of the mRNA molecules for
the pim1 gene is high, then a single strand of the transition molecule displaces
it with an mRNA strand. This displacement process is thermodynamically

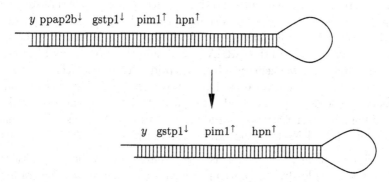

Fig. 7.20 Encoding of a diagnostic rule for the gene mdm2 given by a diagnostic moiety (stem) and a drug-administering moiety (loop). The transition step involves the diagnostic moiety and an active transition rule $y \overset{ppap2b^{\downarrow}}{\longrightarrow} y$.

favorable because of the higher complementarity between transition strand and mRNA. The resulting DNA/mRNA hybrid complex will be degraded in the cell by the endonuclease RNase H, which catalyzes the cleavage of the involved RNA. Hence, the gene will not be translated (Fig. 7.21).

The transition rule $y \overset{pim1^{\uparrow}}{\longrightarrow} y$ is *initially inactive* and therefore not able to interact with the diagnostic moiety. This transition molecule contains a

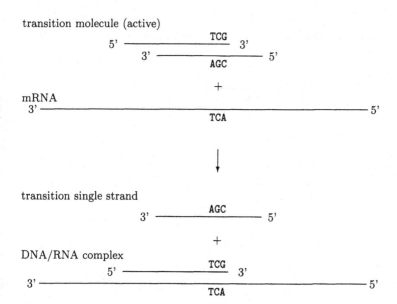

Fig. 7.21 Strand displacement among active transition molecule $y \overset{pim1^{\uparrow}}{\longrightarrow} n$ and mRNA.

mismatch region (internal loop) and there is an oligonucleotide that fully complements one strand of the transition molecule. If the concentration of the mRNA molecules for the pim1 gene is high, then a single strand of the transition molecule displaces the mRNA strand, and the other strand hybridizes with the fully complementary oligonucleotide. This displacement process is thermodynamically favorable, due to the higher complementarity among transition strands, oligonucleotide, and mRNA. The resulting DNA/mRNA hybrid complex is not a viable structure and will be degraded in the cell by the endonuclease RNase H. Moreover, strand displacement provides an active transition molecule $y \xrightarrow{\text{pim1}^\uparrow} y$ that is able to interact with the diagnostic moiety (Fig. 7.22). Ideally, one pim1 mRNA molecule inactivates one transition molecule $y \xrightarrow{\text{pim1}^\uparrow} n$ and activates one transition molecule $y \xrightarrow{\text{pim1}^\uparrow} y$.

An underexpressed gene gstp$^\downarrow$ also gives rise to two transition rules. Now, however, the transition rule $y \xrightarrow{\text{gstp}^\downarrow} y$ is initially active, while the transition rule $y \xrightarrow{\text{gstp}^\downarrow} n$ is initially inactive.

When all symbols are processed, the drug-administering moiety will be processed, too. This moiety consists of a double-stranded stem that contains

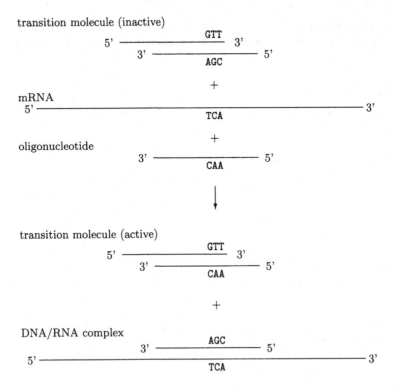

Fig. 7.22 Strand displacement among inactive transition molecule $y \xrightarrow{\text{pim1}^\uparrow} y$, oligonucleotide, and mRNA.

a single-stranded loop, which hosts a drug or drug-suppressor. The loop protects the drug or drug suppressor from unwanted interaction with the target mRNA. In view of a positive diagnosis, the stem of the drug moiety is cleaved and the drug is released, while the stem of the drug suppressor moiety remains intact, protecting the drug suppressor. The situation is reversed in the case of a negative diagnosis.

Each molecular automaton autonomously performs a stochastic computation, and multiple automata operate in parallel carrying out the same task within the same environment and without mutual interference. The behavior of these automata largely depends on the concentrations of the involved molecules and the fidelity of the strand displacement processes. Unfortunately, the specific mechanism proposed would not work in a living cell since unwanted side effects of the FokI enzyme would be a major problem. Nevertheless, the work can be considered to be a big conceptual step forward linking molecular automata to anti-sense technology.

7.5 Computational Genes

Computational genes provide another type of finite state automata, which autonomously operate at the molecular scale. Computational genes invented by the authors (2007) are able to detect and correct aberrant molecular phenotype given by mutated genetic transcripts.

7.5.1 Basic Model

A *computational gene* is a molecular automaton that consists of a structural and a functional moiety and is supposed to work in a cellular environment. The structural part is a naturally occurring gene or operon that is used as a skeleton to encode a drug such as a protein or peptide. The latter comprises the functional part of a computational gene (Fig. 7.23).

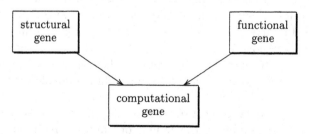

Fig. 7.23 Moieties of computational gene.

In a computational gene, the structural gene translates into a gene product that provides the functional gene. To this end, the conserved features of a structural gene serve as *constants* of a computational gene. These include the promoter of the structural gene, including the RNA polymerase binding site; start and stop codons; and the splicing sites, which control and ensure the smoothness of the splicing process. On the other hand, the coding regions, the number of exons and introns, the position of the start and stop codons, and the automata theoretic variables (symbols, states, and transitions) are the *design parameters* of a computational gene. The constants and design parameters are linked by several logical and biochemical constraints. For instance, encoded automata theoretic variables must not be recognized as splicing junctions.

The input of a computational gene are molecular markers given by single-stranded DNA molecules. These markers signalize aberrant molecular phenotypes, which turn on the self-assembly of the functional gene. If the input is accepted, the output encodes a double-stranded DNA molecule, a functional gene, which should be successfully integrated into the cellular transcription and translation machinery. Otherwise, the output is a partially double-stranded DNA molecule that cannot be recognized by a translation system.

Example 7.20. The computational gene ID1-CDB3 is based on the human inhibitor of DNA binding 1 (ID1) gene as a structural gene that encodes the CDB3 peptide as a functional gene. The ID1 gene (155 aa) is responsible for the production of the ID1 inhibitory protein, which regulates tissue-specific transcription with several cell lines and contributes to cell growth, senescence, differentiation and angiogenesis. It is comprised of two exons (452 bp and 115 bp) separated by an intron (239 bp). The peptide CDB3 (9 aa) can bind to the tumor suppressor protein p53 core domain and stabilize its fold.

In view of the implementation of the computational gene ID1-CDB3, the CDB3 drug given by the amino acid sequence REDEDEIEW-NH$_2$ should be encoded along two exons, but not by the first exon alone. The latter would allow the synthesis of CDB3 without making use of the diagnostic rule (Fig. 7.27). According to the genetic code, R $\in$ {AGA, AGG, CGA, CGC, CGG, CGT}, E $\in$ {GAA, GAG}, D $\in$ {GAC, GAT}, I $\in$ {GAA, GAG}, and W $\in$ {TGG}. Hence, the number of DNA strings to encode CDB3 is given by $6 \times 2 \times 2 \times 2 \times 2 \times 2 \times 3 \times 2 \times 1 = 1152$ encoding sequences. From this pool, those sequences must be eliminated that do not satisfy the necessary logical and biochemical constraints. In particular, a valid DNA encoding sequence must contain the 5'-splicing site region of the first exon (AG) and the 3'-splicing site region of the second exon (G) so that the spliced product of both exons contains the corresponding string AGG in the transcript. Three possible encoding options are shown in Table 7.3.

An encoding sequence is considered to be *optimal* if the sequence satisfies all restrictions and minimizes the number of mutations required to encode the functional gene by the structural gene. Such an optimal string may be found

Table 7.3 Three DNA encoding strings for the CDB3 drug along two exons. Underlined nucleotides describe the CDS region of the first exon and non-underlined nucleotides mark the CDS region of the second exon.

CDB3	W Trp	E Glu	I IIe	E Glu	D Asp	E Glu	D Asp	E Glu	R Arg
1	TGG	GAA	ATA	GAG	GAC	GAA	GAC	GAA	AGA
2	TGG	GAG	ATA	GAG	GAC	GAG	GAC	GAA	CGC
3	TGG	GAA	ATA	GAA	GAC	GAA	GAC	GAA	AGG

by site-directed mutagenesis (e.g., the second encoding string in Table 7.3 requires 17 mutations to encode CDB3 via the ID1 gene (Table 7.4)).   ◇

A computational gene is formally defined as a finite state automaton whose language is given by those double-stranded DNA molecules that are recognized as genes by the translation machinery. In view of Theorem 6.3, linear self-assembly is equivalent to regular languages. Moreover, by Theorem 2.47, regular languages are exactly those languages that are accepted by finite state automata. Therefore, we can expect that computational genes construct functional genes by linear self-assembly. A functional gene can in principle produce any protein without size restriction, as the underlying structural gene is being recognized by the translation system.

7.5.2 Diagnostic Rules

A single disease-related mutation can be diagnosed and treated by the diagnostic rule

$$\text{if gene}X\_\text{mut\_at\_codon\_}Y \text{ then produce\_drug fi} . \qquad (7.11)$$

Table 7.4 Required mutations to encode CDB3 sequence 2 (Table 7.3) by the ID1 gene. Underlined nucleotides represent mutations to be made in ID1. The CDB3 string is enclosed by the start and the stop codon.

Gene/Drug	First Exon String	Second Exon String
ID1	ATCAGCGCCCTGACGGCCGAG	GCGGCATGCGTT
CDB3	ATGTGGGAGATAGAGGACGAG	GACGAACGCTGA

This rule allows the suppression of a pathogenic phenotype and the expression of the wild-type protein or a drug, and thus the restoration of the physiological functionality of the wild-type. The rule (7.11) can be implemented by a two-state, one-symbol automaton. The input symbol corresponds to the mutation in question. If the input symbol is present, the automaton switches over from the initial into the final state. The automaton can be realized by two partially double-stranded DNA molecules and one single-stranded DNA molecule that corresponds to the input symbol. If the mutation is present, the three molecules self-assemble into a double-stranded molecule that encodes the drug (Fig. 7.27).

Example 7.21. For instance, a mutation at codon 249 in the tumor suppressor p53 protein is characteristic for hepatocellular cancer, and the CDB3 peptide can bind to the p53 core domain and stabilize its fold. The corresponding diagnostic rule is given by

$$\text{if p53\_mut\_at\_codon\_249 then produce\_CDB3 fi .} \qquad (7.12)$$

$\Diamond$

The diagnostic rule (7.11) can be generalized so that it corresponds to finite number of disease-related mutations,

$$\text{if geneX\_mut\_at\_codon}Y_1 \wedge \ldots \wedge \text{geneX\_mut\_at\_codon}Y_n \\ \text{then produce\_drug fi .} \qquad (7.13)$$

This rule can be implemented by an $n + 2$-state, n-symbol automaton, where the input symbols are associated with the mutations (Fig. 7.24). If the ith mutation is present, the automaton passes from the $i - 1$-th state into the i-th state. The automaton can be realized by two partially double-stranded DNA molecules, oligonucleotides related one-to-one with the mutations, and further (complementary) oligonucleotides necessary for linear self-assembly. A functional gene will be self-assembled if and only if the n-th (final) state is reached, that is, all n diagnosed mutations are present.

The rule (7.13) may be generalized to involve mutations from different genes, allowing a combined diagnosis and therapy. Furthermore, computational genes are extendable to prokaryotic genes evidencing the generality of the principle. A prokaryotic model could release several different output molecules in response to different environmental conditions, facilitating even more complex computations (Fig. 7.25).

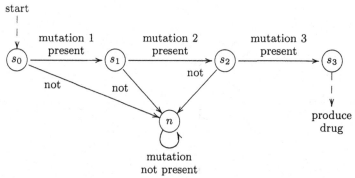

Fig. 7.24 Diagnostic rule implemented by finite state automaton.

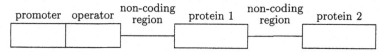

Fig. 7.25 Prokaryotic computational gene.

7.5.3 Diagnosis and Therapy

Computational genes specifically combine the techniques of gene therapy and gene silencing. To see this, observe that in order to process the diagnostic rule (7.11) or (7.13), the molecular automaton must be able to detect point mutations. This task is based on a *diagnostic complex*, a double-stranded DNA molecule, whose single-stranded constituents are termed *mutation signal* and *diagnostic signal*. The signals are not completely complementary to each other: they imperfectly pair in a region that resembles an aberrant mutation to be detected. In this case, an mRNA molecule bearing the corresponding mutation will trigger the dissociation of the diagnostic complex and will perfectly pair with the mutation signal, while the diagnostic signal will be released (Fig. 7.26). This strand displacement process is thermodynamically favorable due to full complementarity between mutation signal and mRNA. The resulting DNA/RNA hybrid complex will be degraded in the cell by the endonuclease RNase H.

The released diagnostic signal provides input to the computational gene and thus contributes to the self-assembly of the functional gene, whose structure is completed by cellular ligase present in both eukaryotic and prokaryotic cells (Figs. 7.27 and 7.28). The transcription and translation machinery of the cell is then in charge of therapy, administering the drug encoded by the functional gene.

Example 7.22. The computational gene ID1-CDB3 can be used to implement the diagnostic rule (7.12). This requires two partially double-stranded molecules, the first of which contains the promoter and the second of which encodes the CDB3 peptide. Here, the diagnostic signal simply provides a

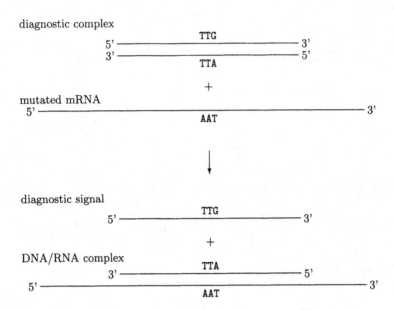

Fig. 7.26 Diagnosis: Diagnostic DNA complex with codon mismatch (internal loop) and mutated mRNA provide a partially double-stranded DNA/RNA molecule given by pairing of mRNA with the mutation signal, while the diagnostic signal is released.

molecular switch that turns on the self-assembly of the functional gene (Fig. 7.27).   ◇

Computational genes might allow the detection of disease-related mutations as soon as they emerge in the cell, and to administer an output that

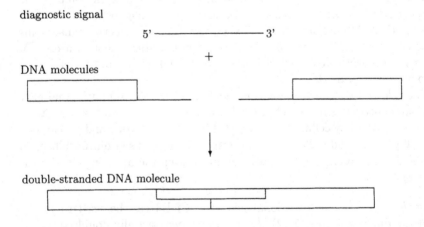

Fig. 7.27 Therapy: The released diagnostic signal activates the spontaneous self-assembly of a double-stranded DNA molecule via annealing and ligation.

diagnostic signals

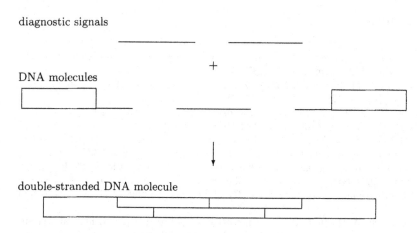

DNA molecules

double-stranded DNA molecule

Fig. 7.28 Therapy: The released diagnostic signals based on two mutations activate the spontaneous self-assembly of a computational gene via annealing and ligation.

simultaneously suppresses the aberrant disease phenotype and restores the lost physiological function. Hence, computational genes might provide therapy in situ when the cell starts developing defective material. In view of the in vivo application of computational genes, the hurdles are similar to those in gene therapy (i.e., the internalization of the computational gene software into the cell, its longevity and stability, and its integrity in the cell).

Concluding Remarks

DNA models for understanding and manipulating cellular behavior are generally invaluable. However, several issues need to be addressed before cellular DNA models can be implemented in vivo. First, the DNA material must be safely internalized into the cell, specifically into the nucleus. In particular, the transfer of DNA or RNA through the biological membranes is a key step in drug delivery. Second, the DNA complexes should have low immunogenicity to guarantee their integrity in the cell and their resistance to cellular nucleases. Current strategies to eliminate nuclease sensitivity include modifications of the oligonucleotide backbone, but along with increased stability, modified oligonucleotides often have altered pharmacological properties. Third, similar to other drugs, DNA complexes could cause non-specific and toxic side effects. In vivo applications of anti-sense oligonucleotides showed that toxicity is largely due to impurities in the oligonucleotide preparation and lack of specificity of the particular sequence used. Undoubtedly, progress in anti-sense technology would also result in a direct benefit to cellular DNA computing.

References

1. Anderson JC, Magliery TJ, Schultz PG (2002) Exploring the limits of codon and anti-codon size. Chem & Biol 9:237–244
2. Alton E (2007) Progress and prospects: gene therapy clinical trials (part 1). Gene Ther 14:1439–1447
3. Bayer T, Smolke C (2005) Programmable ligand-controlled riboregulators of eukaryotic gene expression. Nat Biotech 23:337–343
4. Benenson Y, Gil B, Ben-Dor U, Adar R, Shapiro E (2004) An autonomous molecular computer for logical control of gene expression. Nature 414:430–434
5. Condon A Automata make anti-sense (2004) Nature News Views 429:351–352
6. Drude I, Drombos V, Vauleon S, Müller S (2007) Drugs made of RNA: development and application of engineered RNAs for gene therapy. Mini Rev Med Chem 7:912–931
7. Ehrenfeucht A, Harju T, Petre I, Prescott D, Rozenberg G (2003) Formal systems for gene assembly in ciliates. Theoret Comp Sci 292:199–219
8. Ehrenfeucht A, Harju T, Petre I, Prescott D, Rozenberg G (2004) Computing in Living Cells. Springer, Heidelberg
9. Ehrenfeucht A, Harju T, Petre I, Rozenberg G (2002) Characterizing the micronuclear gene patterns in ciliates. Theory Comput Syst 35:501–519
10. Ehrenfeucht A, Petre I, Prescott D, Rozenberg G (2001) String and graph reduction systems for gene assembly in ciliates. Math Struct Comp Sci 12:113–134
11. Freund R, Martin-Vide C, Mitrana V (2002) On some operations on strings suggested by gene assembly in ciliates. New Gen Comp 20:279–293
12. Ehrenfeucht A, Harju T, Petre I, Rozenberg G (2002) Gene assembly through cyclic graph decomposition. Theoret Comp Sci 281:325–349
13. Harju T, Rozenberg G (2003) Computational processes in living cells: gene assembly in ciliates. LNCS 2450:1–20
14. Harju T, Li C, Petre I, Rozenberg G (2006) Parallelism in gene assembly. Nat Comp 5:203–223
15. Hohsaka TY, Ashizuka H, Taira H, Murakami H, Sisido (2001) Incorporation of non-natural amino acids into proteins by using four-base codons in an E. coli in vitro translation system. Biochem 40:11060–11064
16. Hohsaka TY, Ashizuka H, Murakami H, Sisido (2001) Five-base codon for incorporation of non-natural amino acids into proteins. Nucleic Acids Res 29:3646–3651
17. Kari L, Kari J, Landweber LF (1999) Reversible molecular computation in ciliates. In: Karhum J, Maurer H, Păun, Rozenberg G (eds.) Jewels are Forever. Springer, Heidelberg 353–363
18. Landweber LF, Kari L (1998) The evolution of cellular computing: nature's solution to a computational problem. LNCS 2950:207–216
19. Landweber LF, Kari L, (1999) Universal molecular computation in ciliates. In: Landweber LF, Winfree E (eds.) Evolution as computation. Springer, Heidelberg
20. Landweber LF (1999) The evolution of cellular computing. Biol Bull 196:324–326
21. Martinez-Perez I, Ignatova Z, Gong Z, Zimmermann KH (2007) Computational genes: a tool for molecular diagnosis and therapy of aberrant mutational phenotype. BMC Bioinform 8:365
22. Nakagawa H, Sakamoto K, Sakakibara Y (2006) Development of an in vivo computer based on E. coli. LNCS 3892:203–212
23. Prescott D, Rozenberg G (2002) How ciliates manipulate their own DNA. Nat Comp 1:165–183

24. Rinaudo K, Bleris L, Maddamsetti R, Subramanian S, Weiss R, Benenson Y (2007) A universal RNAi-based logic evaluator that operates in mammalian cells. Nat Biotech 25:795–801
25. Salmons B, Gunzburg WH (1993) Targeting of retroviral vectors for gene therapy. Human Gene Ther 4:129–141
26. Tamm I, Wagner M (2006) Anti-sense therapy in clinical oncology: preclinical and clinical experiences. Mol Biotechnol 33:221–238

Index